Chemical Engineering
for Chemists

Chemical Engineering
for Chemists

Richard G. Griskey
Stevens Institute of Technology

ACS Professional Reference Book

American Chemical Society
Washington, DC

Library of Congress Cataloging-in-Publication Data

Griskey, Richard G., 1931–

   Chemical engineering for chemists/
Richard G. Griskey.

   p. cm.—(ACS professional reference
book)

   Includes bibliographical references (p.
   –   ) and index.

ISBN 0–8412–2215–0

1. Chemical engineering.

I. Title. II. Series.

TP145.G78   1996
660—dc20   96–21026
                            CIP

Photo on cover courtesy of Safety-Kleen Corp.

About the Author

RICHARD G. GRISKEY, Institute Professor Emeritus of Chemistry and Chemical Engineering at Stevens Institute of Technology since 1992, was born in Pittsburgh, Pennsylvania. He received his B.S. in chemical engineering from Carnegie-Mellon University in 1951. He served in the U.S. Army Combat Engineers from 1951 to 1953 (19 months in the Far East during the Korean War). He was then awarded an M.S. and a Ph.D. from Carnegie-Mellon.

Griskey has served as a professor and administrator at many institutions, including the University of Cincinnati, Virginia Polytechnic Institute and State University, the University of Denver, New Jersey Institute of Technology, the University of Wisconsin—Milwaukee, the University of Alabama—Huntsville, Stevens Institute of Technology, and Vanderbilt University. Griskey has held every academic rank (including distinguished professor); he has been a department chair, dean of engineering, provost, and executive vice-president.

Griskey is a fellow of the American Institute of Chemical Engineers (AIChE), the American Society of Mechanical Engineers, and the American Institute of Chemists. He received an Exceptional Achievement Award from the American Chemical Society (Division of Industrial and Engineering Chemistry, Inc.) and has been nominated five times for AIChE's Alan P. Colburn Award (awarded annually to authors of technical papers). He is a member of Sigma Xi and Tau Beta Pi. He has written more than 200 publications, several monographs, and three textbooks. He is listed in *Who's Who in America* and *Who's Who in Engineering*.

Griskey has held visiting professorships at the University of São Paulo in Brazil, Monash University in Australia, and the Institut du Pétrole in Algeria and has been an invited lecturer for the Royal Society of Chemistry in England. In 1971, the National Academy of Science appointed him senior visiting scientist to Poland.

Griskey has full-time industrial experience with Gulf Oil Research, DuPont, and Celanese. Consulting experience includes Hoechst-Celanese, Phillips Petroleum, Monsanto, 3M, ICI, Hewlett-Packard, American Cyanamid, and Litton Industries.

To my wife Pauline,
my children David and Paula,
my mother Emma,
and my late father George

Contents

Preface

The impact of chemistry on the world is immense. Its importance is realized in basic chemical and petroleum industries, as well as in high-technology areas such as energy, environmental, biochemical, biomedical, and advanced materials, which are rapidly shaping the future. Yet although it is obvious that chemistry is a necessary ingredient in our world, it has been less clear in academic circles what skills are needed by scientists and engineers employed in the chemical industry and government to be successful professional practitioners.

To better understand the problem of educational needs, let us first consider chemistry to be a continuum that ranges from theoretical science to the application of chemical principles in technology. A simplistic view of a professional's responsibility within the continuum would be that a chemist takes care of theory and a chemical engineer handles application. Unfortunately, this is not how the industrial world works: Individual chemists and chemical engineers often find themselves at various places in the continuum. Chemists' responsibilities many times include applications, and chemical engineers find that they require a sound knowledge of basic chemistry. Hence, it is apparent that the practicing chemical professional in industry and government requires a broad range of chemical knowledge.

Academic training of chemical engineers already acknowledges the need for a broad chemical education because it mandates that undergraduate students take significant course work in analytical, organic, and physical chemistry, all, incidentally, usually taught by the chemistry department at colleges and universities. Degree accreditation, which is a joint effort by the American Institute of Chemical Engineers (AIChE) and the Accreditation Board for Engineering and Technology (ABET), is not granted unless the chemistry requirements are met. Actually, the total credits of chemistry required by the chemical engineering degree far exceed the number normally required for a chemistry minor in most liberal arts programs. The additional applied chemistry courses, such as thermodynamics and kinetics, that are taken by the engineering students make it even more evident that the holder of a baccalaureate in chemical engineering has an exceptionally strong chemistry background.

The inverse, that the academic training of chemists requires a significant number of courses in chemical engineering, is not the case except in certain countries in continental Europe. The lack of engineering courses causes problems for the chemist when he or she ends up in an applications area of the chemistry continuum.

However, the problem can be alleviated by providing chemical professionals with sufficient chemical engineering background, knowledge, and wherewithal that enable them to realize their full potential.

THE APPROACH OF THIS BOOK

Chemists already have a theoretical grounding in many topics that engineers are trained to apply, so the approach of this book is to start with what chemists know and add relevant principles to them. I do not seek to teach a new approach to a skill that chemists already possess. Chemists, for example, learn stoichiometry and certain aspects of thermodynamics, chemical equilibrium, and chemical reaction kinetics. I have chosen to concentrate on new principles and move them from the theoretical to the semiempirical to the empirical approaches needed to use them.

I have selected my approach based on my past experience gained from teaching chemical engineering to chemists. Some people might argue that chemists should take all of the chemical engineering courses they have not had, but I believe that chemists already have a strong background in many of the topics covered in introductory chemical engineering courses. I did not always believe this. While at the University of Denver in the late 1960s, I initiated and taught a graduate course titled "Chemical Engineering for Scientists". One part of the course called for me to teach the students how to do mass balances. I handed out a set of homework problems of the sort that are typical in an engineering course, requiring them to be done by the next class.

At the next meeting, I collected the homework and began to go over the problems. As I did so, the students in the class wore perplexed expressions that changed to surprise when I obtained the correct answer. Sensing that something was out of order, I asked the group what was wrong. Their collective response was, "We've never seen anyone do a mass balance that way." I asked one of them to work the problem at the board, upon which *I* looked perplexed and surprised when the student got the correct answer.

The obvious lesson I learned is that chemists, as do scientists and engineers, thoroughly understand basic mass balances and do not have to learn stoichiometry. Similarly, other topics such as chemical equilibrium are understood in much the same way. Therefore, this text, while it touches on these subjects, will do so only as a brief review and exposure to the chemical engineering approach.

HOW THIS TEXT CAN HELP YOU

The selection of topics used in this book was derived from my experience working with chemists in industry, from my teaching and research activities, and, last but not least, from my interactions with more than 4000 professionals who have taken my three-day course, Chemical Engineering and Process Fundamentals for Chemists, sponsored by the American Chemical Society. The last experience has been especially useful in shaping the text and has provided insight as to how powerful chemical engineering knowledge can be in enhancing one's career. A few examples reported by some of the course alumni will demonstrate this point:

- One individual who came to the class was the supervisor of a large number of professionals (half were chemists and half engineers). The engineers, the supervisor said, were using the supervisor's lack of process knowledge to their advantage, using engineering jargon during meetings and generally making things difficult. After taking the course, the supervisor was able to understand what the engineers were discussing and was, in fact, able to ask more perceptive and applicable questions and to control meetings with more authority. Understanding engineering basics leveled the playing field.
- Another class attendee was a researcher who had large amounts of interesting data that had defied interpretation. A knowledge of how to apply dimensionless groups of numbers to data from different scales of operations (a technique commonly used by chemical engineers) gave this individual the means to both comprehend the significance of the data and the ability to write three refereed journal articles.
- A third person was transferred to a pilot plant group after taking the course. This individual said that the understanding of chemical engineering basics (i.e., heat transfer and reactor design) made it possible to perform on a level orders of magnitude above the precourse level.

There is general concensus among course attendees that they were able to handle a wide range of process situations and problems. Furthermore, everyone agreed that technical communication and jargon problems disappeared.

HOW TO USE THIS BOOK

Each chapter contains examples of problems and their complete solutions, the complexity of which should not be a problem to any chemist, scientist, or engineer who has had calculus. (Most of the problems require more algebra than calculus.) And finally, references and further reading sources are provided at the end of each chapter.

I suggest that you read Chapters 1 through 6 in order because each subsequent chapter builds on the concepts of the previous one. Working through the solved examples in each chapter is extremely helpful. Above all, try to develop a physical sense of the meaning of the materials.

Finally, this text is just a start, not an end. There is a Chinese proverb that says, "A journey of a thousand miles begins with a single step." It is my hope that this text will provide many steps along the journey toward an understanding of chemical engineering principles and applications, a journey that will bring you to an even higher level of professional competence.

RICHARD GRISKEY
88 Pine Grove Avenue
Summit, NJ 07901

Introduction to Chemical Engineering Principles

INTERFACE BETWEEN CHEMISTRY AND CHEMICAL ENGINEERING

Chemistry and chemical engineering, though taught as separate disciplines, are parts of the fabric that is chemical science and technology. Although they are closely related, chemistry and chemical engineering do differ.

Chemistry came first in the cultures of many nations from the earliest times. Chemical engineering came later when the large, complex production during the Industrial Revolution (18th and 19th centuries) demanded skills beyond those offered by chemistry or traditional engineering education. Fluids, previously transported on a small scale in buckets or flasks, needed to be moved by piping and pumps. Heat could no longer be supplied by throwing another log on the fire. Industrial production needed the help of an engineer.

At first, a mechanical engineer seemed to have the knowledge to make large-scale processes possible. But mechanical engineering could not provide the technology needed because it was divorced from chemistry. Mechanical engineers worked with mass. The mole was a foreign concept to them, and molecular weight was as close as they came to chemistry. There was a chasm between chemists and mechanical engineers.

The chemical industry needed engineers familiar with chemistry and production, so chemical engineering was born. At first, chemical engineering was a branch of chemistry, just as organic, physical, inorganic, and analytical chemistry were established. During the Industrial Revolution, however, chemistry was undergoing a major change, developing into a science in which it was necessary to know the whys as much as the hows. Chemistry moved toward elucidating fundamentals and away from applications.

Today chemistry continues to move toward fundamentals, and this development puts chemists at a disadvantage in dealing with industrial processes. The physical chemistry course is an example of chemistry's concentration on fundamentals. Less than 20 years ago, it included treatment of the laws of thermodynamics, solution behavior, phase equilibrium, chemical kinetics, and diffusion phenomena, all applicable to designing and operating industrial processes. In many universities today, however, physical chemistry is dedicated mainly to statistical and quantum mechanics. Applied chemistry is now taught mainly in the engineering curriculum.

WHAT'S IN THIS BOOK

We're in another revolution now. New materials, new industries, and changing economics demand engineering skills from chemists, scientists, and researchers uneducated in engineering. With the proliferation of small technical companies, many classically trained scientists wear the hats not only of the discoverer of a new material, but also of the manager in charge of producing it commercially.

If you're reading this book, you've realized you need some practical engineering skills. This chapter summarizes the principal areas in chemical engineering that will enable you to understand and work successfully in a chemical plant, pilot plant, product development group, or scale-up project.

It is divided into chapters that concentrate on applied thermodynamics, fluid flow, heat transfer, mass transfer, chemical engineering kinetics, process design and control, and engineering economics. Each chapter presents a discussion of fundamentals with a large number of examples and their solutions. The approach should give you a firm grasp of basic principles and the ability to solve practical problems.

MASS BALANCES

An initial course in chemical engineering usually covers mass balances, familiar to chemists as stoichiometry. A mass balance represents nothing more than the Law of Mass Conservation, i.e., what goes into a system must come out or accumulate. On an industrial scale, mass is balanced across entire processes. It is one of the first tasks an engineer undertakes in designing or optimizing a process.

To obtain a mass balance, apply the following systematic approach:

1. Make a schematic flow diagram of the process, showing all process streams flowing into and out of the system, being sure to include recycled and side streams.
2. Note all of the physical and chemical changes taking place, and list all available data about the streams, including flow rates, temperatures, and pressures.
3. Choose a basis for the mass balance calculation (moles or mass) and a time period over which the balance will be calculated.
4. Select the process segment of interest and circle it to establish the streams entering and leaving. This circle is called the *boundary*. Process streams that do not cut the boundary, including recycled flows, are not considered in the mass balance calculations.
5. Develop equations that describe relationships between entering and exiting streams and enable you to calculate the mass balances across the schematic.

You may need to develop several independent equations, one for each unknown variable. Sometimes in complex systems, many components are interrelated, greatly reducing the number of independent equations.

6. To solve the mass balance, track a compound or compounds through the chosen process segment. If a chemical reaction occurs, track an element or a radical.

Solutions for mass balance problems cannot be generalized and are handled by understanding the process. Because of the lack of generalization possible, this book does not devote a chapter to mass balancing. Instead, the mass balances shown in Examples 1.1 and 1.2 will get you started, and special situations in mass balancing will be noted in chapter examples hereafter.

Example 1.1 demonstrates the principle of mass balance, the interaction of input, output, and accumulation.

Example 1.2 illustrates several important mass balance techniques. The first technique shows how to track specific components through a process, rather than characterizing the overall mass flow. The second technique shows how to define the boundary for a mass balance; only those process streams within the boundary are considered in the mass balance calculations. Furthermore, we'll see that you can define as many boundaries as you need to solve the balance of a given situation.

EXAMPLE 1.1. MASS BALANCE ACROSS A STORAGE TANK

A half-full 60,000-gallon gasoline storage tank is filled from five sources in one day. These sources supply 8,000, 7,000, 5,000, 13,000, and 19,000 gallons, respectively. During the day, 57,000 gallons of gasoline are withdrawn. What is the final volume of gasoline in the tank?

The mass balance is solved by following the six steps previously described. The schematic for the first step is a sketch of a tank showing five input and one output streams. No chemical changes take place, only changes in flow. The mass balance is in gallons and the time interval is one day.

The law of conservation is

$$\Sigma m_{\text{input}} = \Sigma m_{\text{output}} + \Sigma m_{\text{accumulation}}$$

for which m is the mass or moles. (When there is no accumulation of flow, the system is operating under *steady-state* conditions. The accumulation here represents unsteady-state conditions.) The basis for this calculation is mass. To convert the units in gallons to mass, we apply

$$m = V\rho$$

where V is volume and ρ is the density of gasoline in weight per unit volume. The mass balance can then be stated:

$$\rho_{\text{input}}\Sigma V_{\text{input}} = \rho_{\text{output}}\Sigma V_{\text{output}} + \rho_{\text{accumulation}}\Sigma V_{\text{accumulation}}$$

If

$$\rho_{\text{input}} = \rho_{\text{output}} = \rho_{\text{accumulation}}$$

then,

$$\Sigma V_{\text{input}} = \Sigma V_{\text{output}} + \Sigma V_{\text{accumulation}}$$

There is no change in the density of the gasoline, so we can simplify our calculation and use volume. Substituting the appropriate values in the balance,

$$(8{,}000 + 7{,}000 + 5{,}000 + 13{,}000 + 19{,}000) \text{ gallons} = 57{,}000 + V_{accumulation}$$

$$V_{accumulation} = -5{,}000 \text{ gallons per day}$$

$$\text{final volume in tank} = (30{,}000 - 5{,}000) = 25{,}000 \text{ gallons}$$

EXAMPLE 1.2. APPLYING MULTIPLE BOUNDARIES

Gas containing sulfur trioxide (SO_3) is piped into the bottom of an absorber column and stripped of SO_3 by contact with 99 wt % sulfuric acid fed into the top of the absorber. The absorbing liquid stream, 99.5 wt % sulfuric acid, exits from the bottom of the absorber and is diluted with water to produce 99 wt % sulfuric acid. A portion of the acid is removed as product, and the remainder is recycled back to the absorber. This system operates with no accumulation.

If 10^5 kg of 99% sulfuric acid are to be produced per day, determine (1) the water flow rate in kilograms needed for the dilution and (2) the kilograms of 99% acid recycled to the absorber per day.

The schematic flow diagram of the process is shown in Figure 1.1a. The basis of our calculations is mass and a period of one day. To determine the dilution water rate, we need to draw a boundary around the process cutting across four streams (Fig. 1.1b): gas entering the absorber, stripped gas leaving the absorber, product acid, and dilution water. The recycled stream is totally within the boundary and so does not enter into the mass balance calculation.

Make hydrogen the basis of our mass balance:

$$H_2 \text{ in entering gas} + H_2 \text{ in dilution water} = H_2 \text{ in exit gas} + H_2 \text{ in product acid}$$

Neither the entering nor exiting gas contains H_2, so these terms drop from the mass balance calculation. The product acid contains both acid and water. Defining the unknown water flow rate as x and substituting the values we know in the equations results in

$$(2/18)(x) = (2/98)(0.99)(10^5 \text{ kg}) + (2/18)(0.01)(10^5 \text{ kg})$$

$$x = (18/98)(0.99)(10^5 \text{ kg}) + (0.01)(10^5 \text{ kg})$$

$$x = 19{,}200 \text{ kg}$$

On a rate basis, the process needs 19,200 kg of dilution water per day.

To calculate the flow rate of the recycle stream, we need to draw another boundary (Figure 1.1c). This new boundary cuts across four streams: acid with absorbed SO_3 (99.5% H_2SO_4), dilution water, product acid (99%), and recycled acid (99%). The flow rates of the recycled and 99.5% H_2SO_4 streams are completely unknown, so we need two independent equations that we can solve simultaneously.

Of an overall basis, we can balance the total kilograms:

$$y + 19{,}200 \text{ kg water} = 10{,}000 \text{ kg product acid} + z$$

where y equals the mass of acid with absorbed SO_3 and z equals the mass of recycled acid.

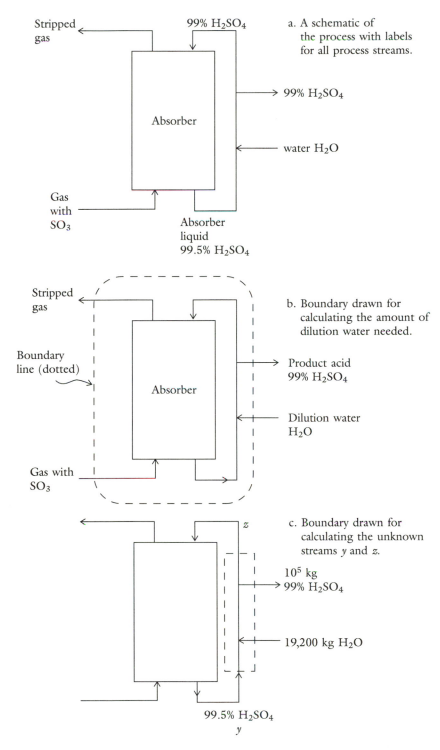

FIGURE 1.1. The first step of a mass balance is to (a) sketch the process, including descriptions of all the process streams. Then define the boundary (b, dotted line), which will cross all the process streams included in the mass balance. The boundary may need to be redefined (c) to balance the mass of other process streams.

To create a second equation, look at the process on a water basis:

$$0.005y + 19{,}200 \text{ kg water} = (0.01)(10{,}000 \text{ kg product acid}) + 0.01z$$

Solving the equations simultaneously yields

$$z = 3{,}720{,}000 \text{ kg recycled acid}$$

or a recycle rate of 3,720,000 kg per day.

ENERGY BALANCES

An energy balance is the counterpart to the mass balance:

$$\text{energy in} = \text{energy out} + \text{energy accumulation}$$

When setting up an energy balance, all forms of energy (potential, kinetic, internal, chemical, vibrational, etc.) must be considered. An energy balance is a form of the first law of thermodynamics, the conservation of energy, which is one of the first topics covered in Chapter 2 on applied thermodynamics. The remainder of Chapter 2 is devoted to using thermodynamic data for flow systems, not generally covered in a chemistry thermodynamics course (the nature of fluid flows is discussed in Chapter 3), real gases, and multiple phases, to prepare for learning about heat transfer in Chapter 4.

HEAT TRANSFER IN INDUSTRIAL CHEMICAL PROCESSES

Large-scale industrial processes operate over wide ranges of temperatures and pressures, creating significant challenges in designing commercially viable, continuously running chemical plants. Temperature changes take place during heating or cooling, as well as in processes that involve phase changes, such as solidification, melting, vaporization, and condensation. In today's competitive economic environment, efficient handling and control of heat transfer lowers production costs and may determine whether a new compound will be commercially viable.

Chapter 4 (Heat Transfer) covers methods for estimating the heat exchange needed for a process, the mechanics of heat transfer (including the modes of convection, conduction, and radiation), and methods for calculating the effects of materials and shapes on heat exchange in reactors and piping.

Although an engineer may be concerned with all types of heat transfer, nowhere is a knowledge of heat transfer more important than when designing or improving reactor efficiency. Because chemists are not generally familiar with the importance of heat transfer in industrial processes, a brief look at temperature and heat transfer effects is helpful before covering the topics in detail.

TEMPERATURE EFFECT ON REACTION RATES

In a chemical reactor, regardless of its configuration or size, the two principal variables affecting the reaction rate are time and temperature. By controlling the heat transfer, and thus the temperature, we can determine the length of time a reaction or process requires for completion.

As a rule of thumb, the rate of a reaction doubles for every 10 °C increase in temperature. To see why the temperature has so much influence, look at the rate of reaction, represented in words by the following equation:

$$\text{rate of reaction} = (\text{specific rate of reaction})(\text{concentration of reactant raised exponentially}) \tag{1.1}$$

or

$$r = k_r C^n_{\text{reactant}} \tag{1.2}$$

where r is rate of chemical reaction, k_r is specific reaction rate or the rate constant, C_{reactant} is concentration of reactant, and n is order of the reaction (exponential). Temperature is important because it causes the specific reaction rate k_r to change, as given by the Arrhenius equation:

$$k_r = Ae^{-E/RT} \tag{1.3}$$

where T is absolute temperature in kelvins, E is activation energy, R is gas constant, and A is frequency. The variation of k_r with temperature is shown in Figure 1.2, which represents typical data one might gather from a rate experiment, where T and k_r are known, and the slope $-E/R$ is plotted.

EFFECT OF HEAT TRANSFER ON REACTION COMPLETION

Heat transfer determines both the operating temperature and the time required for a reaction to reach the desired level of completion. The reactor temperature will vary with the rate of heat transfer, as shown by the experimental data for a batch reactor

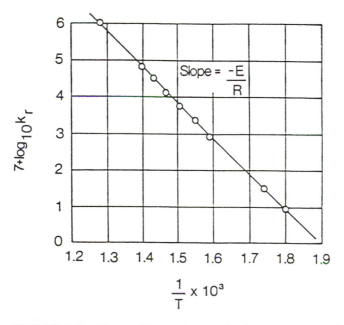

FIGURE 1.2. The specific reaction rate k_r changes exponentially with a change in temperature T as shown by data from a typical rate experiment. Such rapid change is one reason that temperature and heat transfer in an industrial process must be carefully controlled.

in Figure 1.3. A reaction in a batch reactor is analogous to a reaction in a laboratory flask. The reactants are added to the reactor, heated or cooled to control the
reaction, and the products are removed after a predetermined period of time. There
is no flow into or out of the reactor during the reaction.

In an exothermic reaction, when no heat is allowed into or out of the reactor
(adiabatic conditions), the reaction goes nearly to completion in the shortest period
of time possible, and the temperature rises rapidly. If heat is removed at the rate it
is generated (isothermal conditions), reaction completion takes about three times
longer than the adiabatic case. Finally, if heat is removed much more rapidly than it
is generated (i.e., 250 Btu/h removal), the temperature drops rapidly, the reaction
proceeds slowly, and the time required for a reasonable level of reaction completion
becomes impractical.

From Figure 1.3, it appears that we could eliminate many heat transfer problems
by operating a reactor adiabatically, in other words, have no heat transfer whatsoever.
However, adibatic operation would be possible only if four conditions are met:

1. The heat of reaction is small, so that the temperature rise or fall
   is manageable.
2. The initial temperature can be adjusted so that the temperature change
   doesn't take the system out of the reaction range.

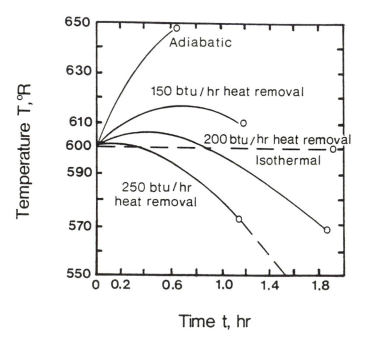

FIGURE 1.3. Controlling heat transfer is critical to smooth and predictable process operation. The rate of heat transfer can determine the time a reaction takes to go to completion and
the temperature of the reaction. The graph shown is for an exothermic reaction under various
rates of heat transfer. If no heat transfer takes place (top curve), the reaction proceeds rapidly
and at ever higher temperatures. If heat transfer is at a rate much greater than the heat generated by the reaction, the temperature drops and the reaction may not reach completion (bottom curve). (Reproduced with permission from Walas et al. Copyright 1989 Walas.)

3. The heat transfer capacity of the equipment, solvent, or inert substances can control the temperature.

4. An inert substance can be used to control the temperature by vaporization or condensation.

THE NONISOTHERMAL REACTOR AS AN INDUSTRIAL STANDARD

The effect of heat transfer can go unnoticed on a laboratory scale because laboratory equipment, generally flasks or small vessels, has a large surface-to-volume ratio. The surface area available for heat transfer makes it simple to maintain the reaction at an isothermal condition. At the industrial scale, however, a reactor's surface-to-volume ratio is much smaller and makes heat transfer and temperature control more difficult. Because of their size, industrial reactors are often nonisothermal. Therefore, knowledge of heat transfer is critical.

The data needed to design or analyze the operation of a nonisothermal reactor include

1. *specific reaction rate* as a function of temperature; this rate is usually determined by a chemist using a knowledge of kinetics;
2. *thermal data* such as heat capacities, sensible and latent enthalpies, covered by thermodynamics in the chemistry curriculum;
3. *heat of reaction* at a base temperature, also covered in chemistry thermodynamics; and
4. *heat transfer rates*, given a set of process conditions, such as fluid flow rates, agitation method, etc.; the engineer would calculate these values using applied thermodynamics, fluid flow, and heat transfer relationships.

The information on engineering thermodynamics in Chapter 2 and on heat transfer in Chapter 4 will introduce you to methods for calculating heat transfer rates.

HEAT TRANSFER EQUIPMENT

Heat can be transferred by a variety of equipment and means. Figures 1.4 through 1.6 are schematics of a few types of equipment.

In a batch reactor (Figure 1.4), chemicals are added and allowed to react for a given time period. Then, the products and unreacted starting materials are removed. To manage the heat transfer, a batch reactor is built with a jacket, a hollow area between the reactor vessel and an outer shell. The jacket is filled with a circulating heat transfer medium, such as steam or a fluid. For additional heat transfer and reaction efficiency, a batch reactor generally has an agitator and baffles.

A flow reactor (Figure 1.5) consists of tubes surrounded by a shell. Reactants and the heat transfer medium (the medium can be another process stream) flow continuously past each other.

More complex heat transfer takes place in equipment such as the catalytic cracker and regenerator (Figure 1.6). Both solid catalyst and several gas streams are fed and removed from the cracker. The reaction and heat transfer take place simultaneously.

HEAT TRANSFER KNOWLEDGE FOR OPTIMIZATION

The ability to analyze the heat transfer of a nonisothermal reactor is useful in the initial design of equipment and also for improving or altering existing processes. Consider the data from a study of the heat transfer and time relationship of a batch re-

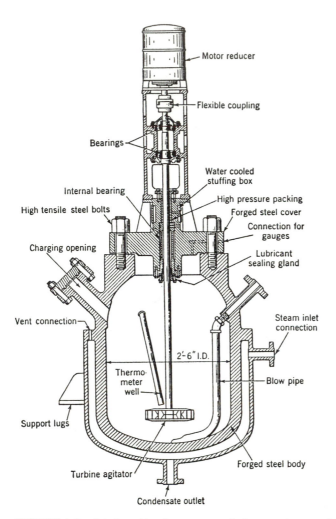

FIGURE 1.4.   A jacketed batch reactor usually has an agitator and baffling for good mixing and a hollow space between the outer shell and the inner reaction vessel for the flow of a heat transfer medium. (Reproduced with permission from Walas et al. Copyright 1989 Walas.)

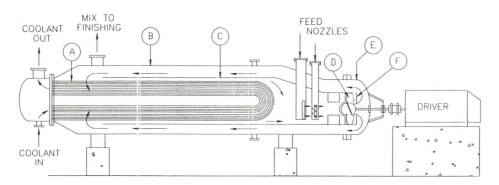

FIGURE 1.5.   A flow reactor, such as this shell-and-tube heat exchanger, has heat transfer medium and process fluid constantly flowing past each other. A, tube bundle; B, shell; C, circulation tube; D, impeller; E, hydraulic head; and F, diffuser vanes. (Courtesy Stratco, Inc.)

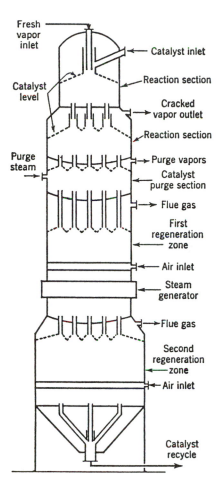

FIGURE 1.6. Calculation of the heat transfer of a process can become quite complex when it combines different phases, as in this catalytic cracker and regenerator. Solid catalyst and process vapor are both fed into the cracker. (Reproduced with permission from Walas et al. Copyright 1989 Walas.)

action, the acetylation of cellulose (Figure 1.7). An analysis of the chemical reactions, thermochemical data, and the heat transfer made it possible to develop a model predicting the effect of an alteration in operating variables. The model led to the conversion of the batch process into a flow reaction; the redesign progressed from small-scale tests to plant-sized operation without any pilot or semiplant testing.

FLUID FLOW AND CHEMICAL PROCESSING

Chemical production requires moving fluids efficiently, rapidly, and safely. Design, operation, and maintenance of process systems involving pumps, fans, valves, piping, and agitators mandate that the chemical practitioner have a thorough knowledge of the mechanics of fluid flow. Chapter 3 covers basic fluid transport and properties, starting with fluid statics (the study of fluids at rest), fluid mechanics (the study of properties that influence flow, including viscosity, density, and pressure), and fluid dynamics (the study of fluids in motion, including the nature and effect of friction, momentum, and velocity profiles).

A second, far-reaching application of fluid flow, as with heat transfer, is its effect on chemical reactor design. To understand why fluid flow analysis is so critical when

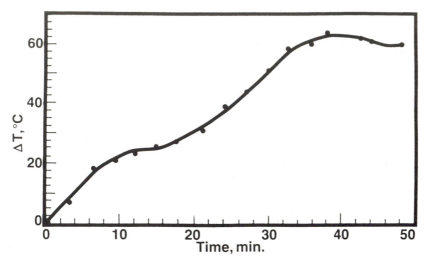

FIGURE 1.7. A history of the temperature and time of an industrial reaction, as shown by this chart for the batch reaction of cellulose acetylation, can be used to model and ultimately optimize the process.

considering reactors, we can review the most common types of reactors found in chemical plants.

In the brief discussion about heat transfer, we introduced the batch reactor (schematic in Figure 1.8), for which the principal operating variables are time and temperature. The temperature, pressure, and composition of the reaction within a batch reactor depend on the process duration (Figure 1.9). The position of a reactant or product within the reactor is not important, provided the contents are well agitated.

Many reactions, however, cannot be conducted in a batch reactor and are, instead, conducted in a flow reactor, so called because the reactants and products pass through the reactor continuously from one end to the other. Reactions requiring a flow reactor include those between gases (control of pressure in a batch reactor would be difficult), rapid or high-volume reactions (the loading and unloading of the batch reactor for such reactions would be inefficient), and reactions that require careful control over concentrations (detailed control and product distribution would be poor in a batch reactor).

Flow reactors have tubular configurations (Figure 1.10) where pressure, temperature, and composition change with position in the reactor. These quantities show

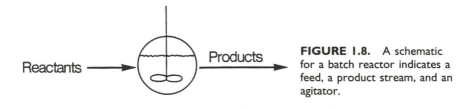

FIGURE 1.8. A schematic for a batch reactor indicates a feed, a product stream, and an agitator.

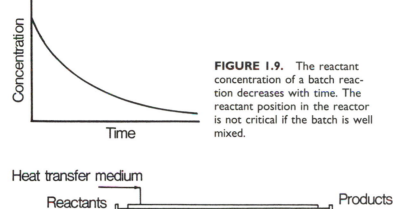

FIGURE 1.9. The reactant concentration of a batch reaction decreases with time. The reactant position in the reactor is not critical if the batch is well mixed.

FIGURE 1.10. The schematic of a tubular flow reactor shows one reactant, one product stream, and the passage of the heat transfer medium in the shell of the reactor. This configuration, with all flows going the same direction, is called cocurrent flow.

no change with time; however, at fixed positions in the reactor length the reactant concentration changes with position in the reactor (Figure 1.11).

A reaction that takes place over a series of continuously charged and stirred tank reactors (each tank is called a stage) (Figure 1.12), shows a similar relationship of concentration and position (Figure 1.13). The semibatch reactor, so called because

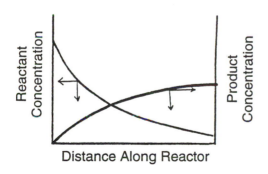

FIGURE 1.11. Reactant and product concentration change with position in a tubular flow reactor.

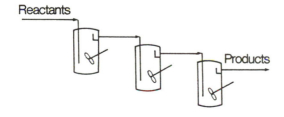

FIGURE 1.12. Schematic of a segment of a continuously charged and stirred tank reactor system.

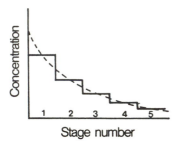

FIGURE 1.13. The reactant concentration in a series of batch reactors shows a profile similar to that of the flow reactor. Concentration varies with the reactor position in the series; each reactor is called a stage.

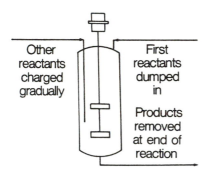

FIGURE 1.14. When a batch reactor is gradually charged with reactants as the reaction takes place, the process is called a semibatch or unsteady-state reaction.

it is a combination of flow and batch reactors (Figure 1.14), exhibits a non-steady-state flow of reactant or product (Figure 1.15).

Knowledge of fluid mechanics is useful in designing and operating flow reactors because the nature of the flow directly affects the reaction. The flow can be affected by the reactor's surface (smooth, rough), the shape of the reactor's cross section (round, square), and the velocity and viscosity of the fluid. Flow can vary from well-mixed without concentration differences in the reactor's tubular cross section to poorly mixed with concentration and velocity differences. Serious concentration differences in the reactor's cross section can result in reduced yields and poor reactor performance.

Chapter 3 covers methods and techniques of handling fluid flow in the situations mentioned above, as well as agitation and mixing, which are special considerations in fluid mechanics.

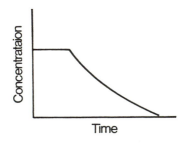

FIGURE 1.15. The reactant concentration profile for a semibatch reaction shows a time dependency after the reactant gradually reaches a reasonable concentration.

MASS TRANSFER IN SEPARATION, PURIFICATION, AND HETEROGENEOUS SYSTEMS

One of the most significant characteristics of today's chemical industry is the heterogeneity of the systems being processed. Compositions, phases, and chemical types vary to yield a vast number of products (see Figures 1.16, 1.17, and 1.18 for examples of interactions between phases). Although the variation of properties greatly increases our production capabilities, it also requires skill in separation and purification. Distillation, absorption, extraction, adsorption, crystallization, filtration, and many other separation processes are ubiquitous throughout the chemical industry, not only for production, but as the principal means of preventing and handling environmental problems. The key to efficient and economical separation, reaction, and purification in all these processes is a knowledge of mass transfer, understanding the nature and control of the preferential movement of chemical species. Mass transfer fundamentals and typical separation and reaction systems are covered in Chapter 5.

EXAMPLES OF MASS TRANSFER

Physical factors related to mass transfer can affect the progress of a reaction even more than chemical affinity. To illustrate how mass transfer can control a reaction,

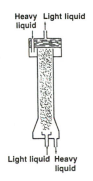

FIGURE 1.16. Making use of density differences is one way of forcing mass transfer between two immiscible liquids. The lighter liquid is fed into the bottom of a spray tower and exits from the top. The heavy liquid fed into the top contacts the lighter liquid and exits from the bottom. The spray tower breaks a fluid into droplets to enhance mass transfer.

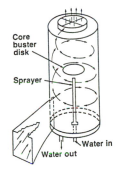

FIGURE 1.17. A scrubber removes undesirable elements in a gas stream by enhancing the contact of water with the gas stream. Mass transfer is enhanced by the breakup of the water into tiny droplets by a sprayer.

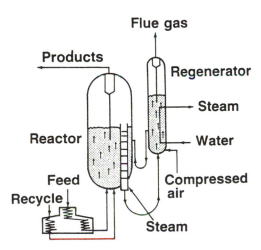

FIGURE 1.18. Mass transfer control improves the reaction in this fluidized bed.

consider a graphic experiment involving the uncatalyzed, heterogeneous reaction of two liquids, aqueous-phase sebacoyl chloride and organic-phase hexamethylene diamine, which results in the formation of a thick white film at the phase interface (Figure 1.19). The reaction takes place only at the interface between the two phases, so the reactants must migrate there. The reaction itself takes place rapidly, limited only by the lack of reactants and removal of the product (nylon film), which can choke off the reaction zone.

The mass transfer steps for the two-phase reaction are (1) the rate of transfer by diffusion or convection of the reactants to and across the interphase boundary, (2) the rate of transfer by diffusion or convection of products away from the reaction zone, and (3) the nature and size of the interfacial area and their relationship to the reaction rate.

Physical mass transfer factors often dominate the course of a reaction. The combustion of coal in air illustrates the importance of mass transfer between a solid and a gas phase. The reaction of coal combustion in air is very rapid, but industrial power plants attain only 1% of the rate possible because oxygen cannot be supplied rapidly enough to the coal surface. The data verify that air velocity is the controlling factor in the reaction (Figure 1.20). Up to about 1100 K, the combustion reaction is solely a function of temperature. At higher temperatures, air velocity becomes a defining parameter.

Ion exchange, a reaction between a liquid and a solid phase, is another illustration of the importance of mass transfer. In ion exchange, the rate of reaction is controlled by (1) the diffusion of ions through the liquid to the resin surface, (2) the diffusion of ions through the resin to the reaction surface, and (3) the chemical reaction rate at the reaction site. In most cases, the chemical reaction is rapid, so again, mass transfer effects dominate.

As a final example, consider the nitration of liquid benzene by concentrated aqueous nitric acid in the presence of sulfuric acid, a catalyst. The reaction is rapid, limited only by lack of contact between the two reacting liquids. Both liquids must be saturated with each other and with the catalyst for the reaction to take place as rapidly as possible.

Successful control and understanding of mass transfer mechanisms is particularly critical when dealing with catalyzed heterogeneous reactions, the principal means of synthesis in the petrochemical and petroleum industries. (Figure 1.21 is the schematic

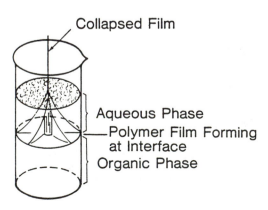

Collapsed Film

Aqueous Phase

Polymer Film Forming at Interface

Organic Phase

FIGURE 1.19. Mass transfer, not the specific chemical reaction rate, determines the overall efficiency of reaction in the laboratory-scale production of nylon from aqueous and organic liquid phases. Lack of fresh reactants and a buildup of nylon at the interface, both problems of mass transfer, slow the reaction.

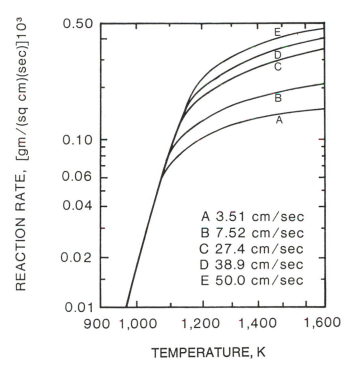

FIGURE 1.20. The speed at which air reaches coal surfaces determines the reaction rate of coal combustion in air above 1100 K. The rate of combustion increases with higher air velocities, as shown by the various curves.

of a petroleum plant reactor in which a process gas is placed in contact with solid catalyst.) The catalysts used are generally solids because they are thermally stable and easy to separate from the reaction mix. When analyzing such heterogeneous reactions, the mass transfer mechanisms are generally broken down into five steps: (1) diffusion of reactants to the catalyst surface, (2) adsorption of reactants on the catalyst surface, (3) reaction on the catalyst surface, (4) desorption of products from the catalyst surface, and (5) diffusion of products into the reaction fluid. If we attempt to link and analyze all of these steps together, we would have a difficult problem. Fortunately, in most mass transfer problems, few, and sometimes only one, of the steps predominate, greatly simplifying an expression for the rate of reaction.

CHEMICAL ENGINEERING KINETICS, PROCESS CONTROL, AND ENGINEERING ECONOMICS

CHEMICAL ENGINEERING KINETICS

Engineering kinetics, the subject of Chapter 6, represents an applied form of chemical kinetics. Emphasis is placed on areas of industrial importance, such as non-isothermal reactions, systems other than batch reactors (flow, continuous stirred tanks, semibatch), heterogeneous reaction systems (involving both uncatalyzed and catalyzed cases), and fixed, moving bed, and fluidized-bed reactors.

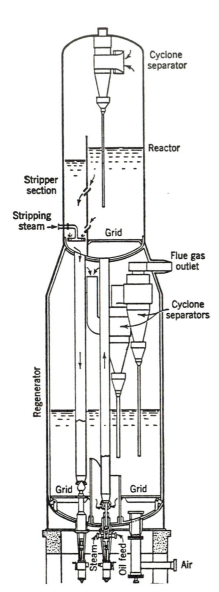

FIGURE 1.21. Mass transfer calculations for a process as complex as the catalytic cracking unit can often be simplified by considering only the dominating steps. (Reproduced from Walas et al. Copyright 1989 Walas.)

PROCESS CONTROL

Process control deals with the basic mechanisms for controlling a process. Chapter 7 introduces various control concepts, the methodology of process dynamics, and the many permutations used for specific processes.

ENGINEERING ECONOMICS

With the principles of thermodynamics, heat and mass transfer, a fluid flow, and chemical kinetics, the chemical practitioner can improve or design a process, but is still missing one key element: economics. Whether a process is economically feasible is one of the chief questions asked during the design of an industrial process. Chapter 8 offers some tools enabling you to evaluate a system economically, including understanding the time value of money, evaluating alternatives, determining or estimating equipment costs, applying price indexing, and examining the process and plant design sequence.

BASIC STEPS FOR SCALING UP A PROCESS DESIGN

In addition to developing an engineering view of the physical processes that affect chemical reactions and processes, the practicing chemist is often confronted with the need to scale up a system. This challenge requires guidelines for the general scale-up procedure, as follows:

1. *Geometric similarity.* Both the bench and commercial process must be the same geometrically. We cannot conduct a laboratory study in a three-necked flask (a batch reactor) and scale it up to a commercial tubular reactor (a flow reactor).
2. *Dynamic similarity.* The velocities, temperatures, and concentrations must be the same at relative points in the laboratory and process units.
3. *Boundary conditions.* We must have essentially the same conditions in the large scale as in the small scale. If our process is adiabatic in the laboratory, it should not be scaled up to an isothermal process.
4. *Scaling factors.* As a check on the similarity between small- and large-scale processes, we can compare the values of dimensionless groups of numbers for each system. These groups represent ratios of forces or effects in a system, called dimensionless because the units in the ratios ultimately cancel each other. Dimensionless groups of numbers exist for evaluating fluid flow, heat transfer, mass transfer, and chemical reaction. For example, the Reynolds number, the ratio of inertial forces to viscous forces in a fluid flow, is an important scaling factor for comparing the character of flow rates. Some of the most useful scaling factors are summarized in Appendix A.

Scaling factors are discussed in the chapters that cover the related physical process and are applied in the chapter example problems.

THE PRINCIPLES OF FLUID FLOW, HEAT TRANSFER, AND MASS TRANSFER

When designing or optimizing a chemical process, we need to define quantitatively three physical aspects of a system: fluid flow, heat transfer, and mass transfer. We'll calculate an overall transfer rate per unit area, called a *flux*, for each of the physical forces. The common mathematical expression of a flux is a system property multiplied by a *gradient* (the change in a driving force, such as temperature, per unit length).

In heat transfer, temperature differences determined for each of the three coordinates, x, y, and z, act as driving forces. The overall heat transfer rate is derived by solving Fourier's law of heat conduction. The flux in the y direction is

$$q'_y = -k \; dT/dy \tag{1.4}$$

where q'_y is the heat flux in the y direction in units of heat per unit time per unit area, k is the thermal conductivity (system property) in units of heat per unit length per unit time per degree, and dT/dy is the temperature gradient in degrees per unit length.

In mass transfer, differences in concentrations (actually, the activities of the components, which, in many instances, reduce to concentrations) in the three coordinate directions are the driving forces. Fick's law of diffusion for mass transfer is an analog to Fourier's law. In the y direction, the flux expression is

$$J'_{A_y} = -D_{AB}\, dC_A/dy \qquad (1.5)$$

where J'_{A_y} is the mass flux of component A in moles per unit time per unit area, D_{AB} is the diffusion coefficient or diffusivity (i.e., system property) of component A in component B in area per hour, and dC_A/dy is the concentration gradient of A in the y direction in moles per unit volume. As you can see, there is great similarity between heat and mass transfer. We'll explore this similarity later in more detail.

Although the third major physical process, fluid flow or momentum, can be described by a mathematical expression similar to heat and mass transfer, it is much more difficult to define quantitatively because it has nine components, not three. (Momentum is a tensor, not a vector.) In its simplest form, momentum can be described by Newton's law of viscosity:

$$\tau_{yx} = -\mu\, dV_x/dy \qquad (1.6)$$

where τ_{yx} is the shear stress exerted in the x direction in units of force per unit area, μ is the viscosity (system property) expressed in units of mass per unit length per unit time, and dV_x/dy is the velocity gradient in units of the reciprocal of time. Quantitative solutions for momentum are usually found by considering only part of the data and simplifying the calculation.

The flux equations for heat and mass transfer each have three components because heat and mass transfer are vectors, that is, each process has both magnitude and direction in the three coordinates. This condition is fortunate because solutions of differential equations for a given geometry involving either heat or mass transfer will be correct for the other transfer process. The analogy helps in process work when it may be impossible to gather data for both heat and mass processes and may, in fact, be the only way that complex process problems can be handled. Only in special cases is fluid flow analogous to heat and mass transfer.

CHEMICAL ENGINEERING PRINCIPLES APPLIED TO NEW TECHNOLOGIES

One of the gratifying, but nonetheless daunting, subjects for practitioners of science and engineering is the constantly changing array of technologies. Some pessimists say that the constant change accelerates the obsolescence of technologists, but new technologies actually represent innovative or creative applications of existing principles, not new principles.

A pertinent example is the science of thermodynamics. The three laws that govern thermodynamics remain unchanged. Applications, however, have proliferated at a bewildering pace. The entire field of space travel has been greatly dependent on applied thermodynamics. Yet, at the time space travel was envisioned, notably by Jules Verne, the laws of thermodynamics were just being developed. Therefore, a

Table 1.1. How Chemical Engineering Principles Relate to New Technologies

Subject	Thermo-dynamics	Fluid Flow	Heat Transfer	Mass Transfer	Chemical Kinetics
Environmental science and technology					
Chemodynamic fate of chemicals	X	X		X	X
Reactive flows in porous media		X		X	X
Biohazard containment			X	X	X
Toxic and hazardous waste control		X		X	X
Plastic recycling				X	
Membrane technology				X	
Advanced materials					
Chemical vapor deposition	X			X	X
Ceramic processing	X		X	X	X
Composite materials			X	X	
Semiconductor production			X	X	
Optical fiber production	X	X		X	
Electronic packaging		X	X		
Polymer processing	X	X	X	X	X
Biochemical and biomedical engineering and biotechnology					
Artificial kidneys				X	
Bioseparations				X	
Chromatographic separations				X	
Controlled-release systems				X	X
Cell separation	X			X	
Membrane technology				X	
Pharmacokinetics				X	X
Recovery of biologicals				X	
Supercritical extraction	X		X	X	

practitioner who combines a clear understanding of principles together with adaptability, flexibility, creativity, and innovation will be able to meet the challenges of a myriad of technological applications.

The idea that applications and not the principles change can be shown in another way. Table 1.1 lists some of the newest technological areas and applicable chemical engineering principles. The information in Chapters 7 and 8, on process control and economics, is applicable to all of the new technologies.

FURTHER READING

Anderson, L. B.; Wentzel, L. A. *Introduction to Chemical Engineering;* McGraw-Hill: New York, 1961, out of print.

Henley, E. J.; Rosen, E. M. *Material and Energy Balance Computations;* John Wiley and Sons: New York, 1969, out of print.

Himmelblau, D. M. *Basic Principles and Calculations in Chemical Engineering*, 5th ed.; Prentice-Hall: Englewood Cliffs, NJ, 1989.

Hougen, O. A.; Watson, K. M. *Chemical Process Principles: Part I, Material and Energy Balances*, 2nd ed.; John Wiley and Sons: New York, 1954.

Peters, M. S. *Elementary Chemical Engineering*, 2nd ed.; McGraw-Hill: New York, 1984.

Schmidt, A. X.; List, H. L. *Material and Energy Balances;* Prentice-Hall: Englewood Cliffs, NJ, 1962.

Shaheen, E. I. *Basic Practice of Chemical Engineering*, 2nd ed.; International Institute of Technology: Joplin, MO, 1983.

Thatcher, C. M. *Fundamentals of Chemical Engineering;* Charles E. Merrill: Columbus, OH, 1962, out of print.

Tyner, M. *Process Engineering Calculations: Material and Energy Balances;* Books on Demand: University Microfilms International, 1960, may be out of print; found as cited in *Books in Print 1991–1992, Authors.*

Walas, S. *Reaction Kinetics for Chemical Engineers;* Butterworth-Heinemann: Woburn, MA, 1989; out of print.

2

Applied Thermodynamics

Thermodynamics, the interaction of energy and mass, receives considerable attention in both the chemistry and chemical engineering curricula. However, the two take different approaches.

The thermodynamics of chemistry concerns itself mostly with closed or control mass systems, which have no movement of mass across the boundary, as well as with ideal gases and solutions.

In contrast, chemical engineering concerns itself with open systems that have both energy and mass crossing the boundaries, systems that are common in industrial processes. Furthermore, engineers work primarily with real, not ideal, gases and solutions under a wide variety of conditions, including nonideal high pressures and temperatures. Phase equilibrium is a significant subject because it is the basis for understanding and designing separation processes that are ubiquitous in the chemical process industries. For that reason, chemical engineers must have a variety of applied techniques to develop, extend, or estimate thermodynamic data (Table 2.1).

This chapter covers applied thermodynamics from a macroscopic viewpoint, which requires many fewer descriptions than the microscopic molecular one. We'll start with a review of thermodynamics as found in a physical chemistry course, adding information as it applies to industrial systems.

TERMS

You'll find the following terms helpful in understanding applied thermodynamics:

- A *system* is the part of the universe or process under study. For example, the gas in a cylinder, water flowing through a turbine, and Freon-12 in a refrigeration unit are all systems.
- A *property* is any observable characteristic of a system. Temperature, pressure, volume, and entropy are all properties.
- An *extensive property* is one that depends on a system's mass and size. Volume, internal energy, and entropy are extensive properties.

TABLE 2.1   Chemistry and Chemical Engineering Curricula

Topic	Chemistry Curriculum	Chemical Engineering Curriculum
Systems	Closed (energy flow only)	Closed and open (mass and energy flows)
Gases	Mainly ideal	Ideal and real
Entropy and enthalpy	Effect of temperature only	Effect of temperature and pressure
Solutions	Knowledge of nonideal	Extensive treatment of nonideal
Phase equilibrium	Basic understanding	Basic understanding; strong emphasis on applications

- An *intensive property* does not depend on the system mass or size. Pressure, temperature, and molal volume are intensive properties.
- The *state* of a system refers to and is determined by its properties at equilibrium.
- A *process* is a change of state determined by a change in properties. If a system's temperature changes, the increase or decrease in temperature is the process.
- A *process is reversible* if both the system and the surroundings can be restored to their initial state at the end of the process.

TEMPERATURE AND HEAT

To explore the topic of energy, we need a definition of temperature and heat. Any pair of systems without an extraordinary barrier between them will generally affect each other. If the effect involves no exchange in mass across the systems' boundaries, the pair approaches a limiting condition called equality of temperature.

The definition of temperature, in turn, leads to the zeroth law of thermodynamics: If two systems are each equal in temperature to a third system (the third system is typically a thermometer), then, they are equal in temperature to each other.

Heat (Q) can be defined as that substance which is transferred between a system and its surroundings (or vice versa) solely on the basis of a temperature difference. Heat can be thought of as energy in transit. The sign for heat, Q, is fixed by convention as negative for an exothermic case and as positive for an endothermic one.

DEFINITION OF WORK

To use applied thermodynamics, we need to define work, W, the effect of the system on the surroundings (or vice versa) expressed as a product of a generalized force and a generalized displacement. Work is a form of energy, but it is not heat, although it can be converted into heat. In most chemical systems work raises or lowers a weight in the surroundings. The mathematical convention used to describe work is that it is positive ($+$) if done on the surroundings by the system, and negative ($-$) if done on the system by the surroundings. Work's magnitude is *path dependent*. The amount of work done between two states can vary with the path taken (Figure 2.1).

CONSERVATION OF ENERGY—THE FIRST LAW OF THERMODYNAMICS

Work is path dependent with one exception. Imagine a perfectly closed system, for example, a gas contained in a cylinder with a weightless, frictionless piston, surrounded by an adiabatic wall (the adiabatic wall prevents any transfer of heat to or from the system's surroundings). When the piston moves, the magnitude of the work performed by or on the system is related only to the beginning and final states of the system and not the path taken. The work performed adiabatically by the system between two states is equal to a change in a system function called the *internal energy*, U. In equation form,

$$U_{final} - U_{initial} = \Delta W_{adiabatic} \tag{2.1}$$

If the closed system is without the adiabatic wall, so that heat is exchanged with the surroundings, then

$$dU = d'Q - d'W \tag{2.2}$$

$$\Delta U = Q - W \tag{2.3}$$

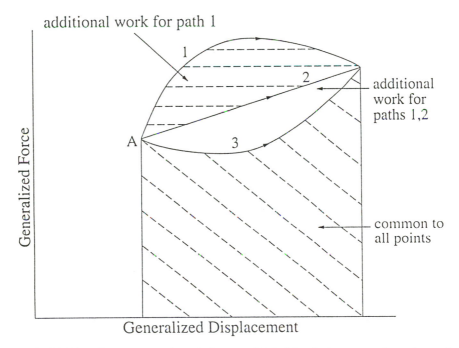

FIGURE 2.1 The quantity of work, force multiplied by displacement, depends on the path taken. Of the paths 1, 2, and 3 between points A and B, path 3 represents the least amount of work, and path 1 the greatest. The lowest area with diagonal stripes is common to all the paths.

in which the primes on heat and work indicate that the differentials are inexact and path dependent. Eq. 2.3 represents the first law of thermodynamics, which is a statement of the conservation of energy: The change of energy in a system must be accounted for by heat or work. The shortcoming of this definition of internal energy is that it does not account for energy related to the gravitational force on objects (potential energy), friction, velocity (kinetic energy), or any other energy outside of heat or a basic force multiplied by a displacement. A more correct statement of the first law is

$$dE = d'Q - d'W \qquad (2.4)$$

where E includes not only U, internal energy, but all other energy forms (potential, kinetic, etc.).

The form of the first law depends on the work mode involved. In chemical processes, the usual form of work is an expansion or compression of a system, calculated by multiplying the pressure, P, of a substance by its volume, V. There are, however, many work modes, including elastic stretching, magnetization, polarization, and variation of the charge of an electrochemical cell. The first law forms for various work modes are shown in Table 2.2.

ENTHALPY

Enthalpy, H, is another thermodynamic function of importance. It is defined for a PV work system under constant pressure (a condition common to chemical laboratory work) as

$$H = U + PV \qquad (2.5)$$

Enthalpy is a point function like internal energy, so that its magnitude is independent of the path followed. Enthalpy is, however, more useful than U because H is operable at constant pressure.

TABLE 2.2 Work Modes and the First Law for Various Work Systems

System	Form of First Law	Properties[a]
Expansion and compression	$dU = d'Q - PdV$	Pressure P, volume V, or temperature T
Stress and strain	$dU = d'Q + \sigma d\epsilon$	Stress σ, strain ϵ, or temperature T
Surface expansion and contraction	$dU = d'Q + \gamma dA$	Surface tension γ, area A, or temperature T
Polarization	$dU = d'Q + EdP$	Electric field strength E, total polarization P, or temperature T
Magnetization	$dU = d'Q + HdM$	Magnetic intensity H, magnetization M, or temperature T
Electrolytic cell	$dU = d'Q + edq$	Electromotive force e, charge q, or temperature T

[a]Internal energy is a function of any two listed properties.

For other work systems, the enthalpy function is given by adding the work expression to internal energy. For a reversible cell, for example,

$$H = U - eq$$

or for a stressed bar,

$$H = U - \sigma\epsilon$$

PHASE BEHAVIOR

To find the amount of work done to or by a system, we have to be able to define the character of the system's phases. The starting point for the definition is the Gibbs phase rule, which determines how many variables must be fixed to describe a given system. (Variables are such items as temperature, pressure, and phase compositions.)

number of coexisting phases + degrees of freedom = number of components + 2

$$P + F = C + 2 \tag{2.6}$$

The phase rule explains the behavior found in plots of pressure and temperature (Figure 2.2) and pressure and volume data (Figure 2.3).

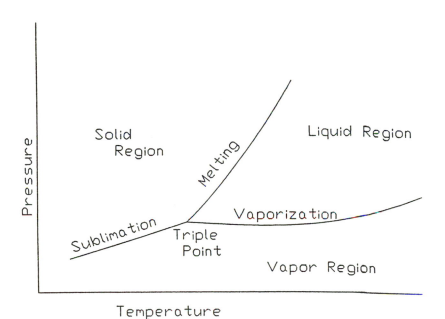

FIGURE 2.2 Thermodynamic properties for a pure substance at equilibrium can be shown by different diagrams. One diagram is a phase diagram, such as this, with pressure vs. temperature. The lines for sublimation, melting, and vaporization are boundaries between two phases; as predicted by the Gibbs phase rule, any point along one of the lines is defined by setting the pressure or temperature. (Adapted from reference 19.)

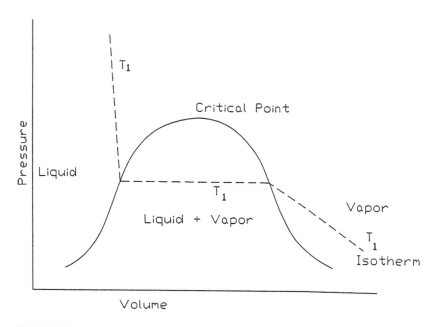

FIGURE 2.3 A thermodynamic diagram for pressure vs. volume defines the phases of a pure substance at different equilibriums. A *PV* diagram consists of a series of isotherms that are represented as vertical lines in a *PT* diagram (Figure 2.2). The area under the dome is a region of liquid and vapor in equilibrium; along a horizontal isotherm from left to right, the equilibrium composition shifts from mostly liquid with a little vapor to mostly vapor with a little liquid. If a vertical line represents a constant volume, at any point along the line, the pressure and temperature of the material are automatically fixed.

Consider, first, the PT diagram for a pure substance (Figure 2.2). How many variables would need to be specified to describe the triple point? At the triple point where three phases coexist (the number of phases is 3), the phase rule would predict

$$3 + F = 3$$

$$F = 0$$

or that no other variable need be specified because there is only one triple point.

In a one-phase (solid, liquid or vapor) region,

$$1 + F = 3$$

$$F = 2$$

two variables, that is, pressure and temperature, must both be specified in order to describe the equilibrium state of the material.

When two phases coexist, as liquid and vapor do at the vapor pressure line,

$$2 + F = 3$$

$$F = 1$$

As soon as we fix the pressure or temperature, the other is automatically established. This is shown by positions on the lines of vaporization, melting, and sublimation, called phase boundaries (Figure 2.2).

A pressure–volume diagram (Figure 2.3) also shows the phase behavior for all of the cases just described. A *PV* diagram consists of a series of isotherms which are represented by vertical lines in Figure 2.2. The region under the dome represents mixtures of liquid and vapor in equilibrium; at the left of the horizontal isotherm, the mixture is more liquid than vapor, and at the right, more vapor than liquid. As soon as we fix the volume, the pressure and temperature are automatically set (by setting the volume, we set the only possible degree of freedom). And if we look at the one-phase region (either liquid or vapor), we must define the values of two variables to fix the thermodynamic state.

ESTIMATING TWO-PHASE THERMODYNAMIC PROPERTIES

In a two-phase region of a pure substance, thermodynamic properties, such as internal energy and enthalpy as well as specific volume, are determined by the *quality* of the substance. The quality is the mass fraction x of the substance in the vapor phase of the two phase system, which we can determine by the lever rule on a pressure-volume plot. Once the quality value is found, the thermodynamic value for the component in the two-phase system is given by the sum of the quality multiplied by the saturated vapor value per mole of the substance and one minus the quality multiplied by the saturated liquid value:

$$\bar{v}_{\text{two phase}} = (x)\bar{v}_{\text{saturated vapor}} + (1-x)\bar{v}_{\text{saturated liquid}} \tag{2.7}$$

$$\bar{h}_{\text{two phase}} = (x)\bar{h}_{\text{saturated vapor}} + (1-x)\bar{h}_{\text{saturated liquid}} \tag{2.8}$$

$$\bar{u}_{\text{two phase}} = (x)\bar{u}_{\text{saturated vapor}} + (1-x)\bar{u}_{\text{saturated liquid}} \tag{2.9}$$

where the lower-case nomenclature and the bar over each property indicate the thermodynamic value per mole (molal value).

EQUATIONS OF STATE

To determine behavior in a thermodynamic system, we need to relate thermodynamic properties, such as pressure, volume, and temperature. This relationship is called an *equation of state*. Equations of state reflect the experimentally determined interrelationships of the properties.

The best known equation of state is the ideal gas law

$$PV = nRT \tag{2.10}$$

where n is the number of moles, R is the gas constant, and T is the absolute temperature. At low pressures (less than 1.013×10^6 N/m^2 or 10 atmospheres), the ideal gas law is a good representation of *PVT* behavior. However, at pressures greater than 10 atmospheres, other equations of state apply.

One such equation of state is the van der Waals equation, which has the general form of the ideal gas equation with corrections for the pressure and volume:

$$(P + a/V^2)(V - b) = nRT \tag{2.11}$$

The term a/V^2 is a correction for intermolecular attractive forces and b is a correction for the size of the molecules. Since a and b can be correlated with the critical point of a substance, we say that the van der Waals equation is based on the Theory of Corresponding States or the reduced state variables. (A reduced state variable, such as P_r, is the ratio of P to its critical point value P_c). The theory says that different real gases will conform to the same equation of state if the reduced properties are substituted for the unreduced ones. The van der Waals equation is an improvement over the ideal gas law equation, but it is not the ultimate for PVT relationships.

An equation of state based on corresponding state behavior that is useful for describing the PVT behavior of real gases is

$$Z = PV/nRT \tag{2.12}$$

or

$$PV = ZnRT \tag{2.13}$$

The compressibility factor, Z, is related to the compound's reduced properties T_r and P_r. T_r is equal to T divided by the critical temperature, T_c, and P_r equals P divided by the critical pressure, P_c. General correlations of Z to the reduced properties are shown in chart form (Figure 2.4). The appropriate value of Z can then be used in eq 2.12.

The compressibility-factor approach to correcting the equation of state can also be used for mixtures of gases. In these cases, the same correlation between Z and the reduced properties is used except that the sum of the critical values, the pseudo-critical value, is substituted in the equation for T and P

$$T_c' = \Sigma y_i (T_c)_i \tag{2.14}$$

and

$$P_c' = \Sigma y_i (P_c)_i \tag{2.15}$$

for which y_i is the mole fraction of gas component i in the mixture, and T_c and P_c are the critical values of the temperature and pressure of component i, respectively.

THERMODYNAMIC FUNCTIONS

There are other thermodynamic functions that simplify the solution of thermodynamic calculations. Specific heat at constant pressure and at constant volume for a PV system are two such functions. Understanding how to estimate the values of thermodynamic properties such as enthalpy, internal energy, and the Joule–Thomson coefficient for real gases, starting from a known base value, is also helpful.

SPECIFIC HEAT AT CONSTANT VOLUME.   For a PV work system, a change in the internal energy is

$$dU = d'Q - PdV \tag{2.16}$$

At constant volume with temperature change,

$$(dU/dT)_v = (d'Q/dT)_v = C_v \qquad (2.17)$$

where C_v is the specific heat at constant volume.

SPECIFIC HEAT AT CONSTANT PRESSURE.   For a change in enthalpy,

$$dH = dU + PdV + VdP \qquad (2.18)$$

Because

$$d'Q = dU + PdV \qquad (2.19)$$

then,

$$dH = d'Q + VdP \qquad (2.20)$$

Therefore, for a temperature change at constant pressure,

$$(dH/dT)_P = (d'Q/dT)_P = C_p \qquad (2.21)$$

where C_p is the specific heat at constant pressure.

SPECIFIC HEATS FOR OTHER WORK MODES.   In the same way that the specific heats were derived for the PV work mode, counterpart specific heats for other work modes can be derived. The change of internal energy for a reversible cell is

$$dU = d'Q + edq$$

and

$$(dU/dT)_q = (d'Q/dT)_q = C_q$$

where C_q is the specific heat at constant charge, and

$$(dH/dT)_e = (d'Q/dT)_e = C_e$$

where C_e is the specific heat at constant electromagnetic force.

RELATIONSHIPS BETWEEN THERMODYNAMIC FUNCTIONS FOR AN IDEAL GAS.   Calculations of the thermodynamic values of ideal gases are staightforward. C_p and C_v are related by

$$\bar{C}_p - \bar{C}_v = -T \left(\frac{\partial \bar{v}}{\partial T} \right)_P^2 \bigg/ \left(\frac{\partial \bar{v}}{\partial P} \right)_T$$

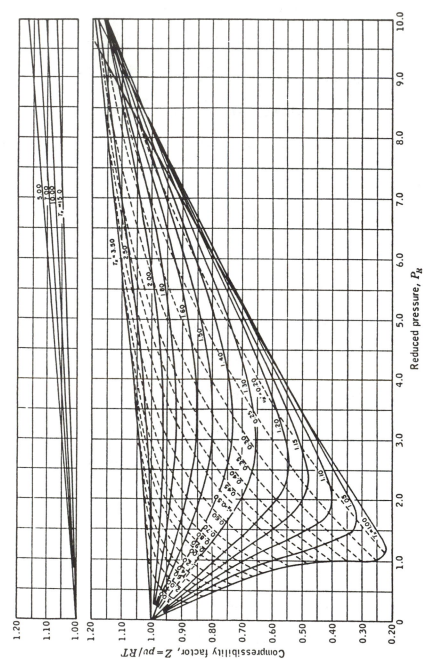

FIGURE 2.4 Behavior of a nonideal gas can be estimated by correcting the ideal gas law with a compressibility factor Z. Z can be estimated from generalized compressibility factors charts, such as these, by first calculating the reduced pressure P_r and reduced temperature T_r of the compound. (Reproduced with permission from reference 1. Copyright 1954 McGraw-Hill.)

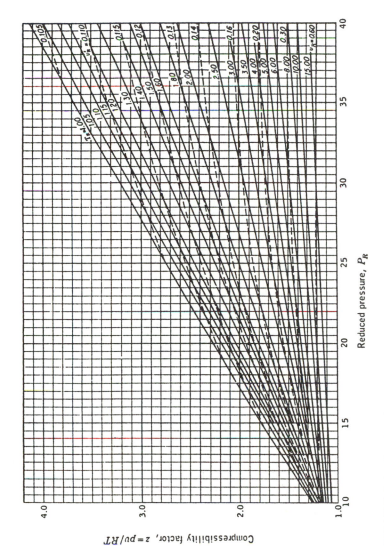

FIGURE 2.4 Continued.

For an ideal gas,

$$\left(\frac{\partial \bar{v}}{\partial P}\right)_T = -\frac{RT}{P^2}$$

$$\left(\frac{\partial \bar{v}}{\partial T}\right)_P = \frac{R}{P}$$

so that the molar values of the specific heats are related by

$$\bar{C}_p - \bar{C}_v = R \tag{2.22a}$$

The enthalpy change due to pressure for a gas is

$$\int_{P_1}^{P_2} \left[\bar{v} - T\left(\frac{\partial \bar{v}}{\partial T}\right)_P \right] dP \tag{2.22b}$$

But for an ideal gas,

$$T\left(\frac{\partial \bar{v}}{\partial T}\right)_P = T\left(\frac{R}{P}\right) \tag{2.23}$$

then,

$$\left[\bar{v} - T\left(\frac{\partial \bar{v}}{\partial T}\right)_P \right] = \left[R\left(\frac{T}{P}\right) - T\left(\frac{R}{P}\right) \right] = 0 \tag{2.24}$$

In other words, for an ideal gas the enthalpy change due to pressure change is zero and the internal energy, U, and enthalpy, H, are functions only of temperature.

In addition, the Joule–Thomson coefficient μ is zero for an ideal gas

$$\mu = (\partial T / \partial P)_H = 0 \tag{2.25}$$

COMPUTING THERMODYNAMIC FUNCTIONS FOR REAL GASES

For real gases, unlike ideal ones, the functions of internal energy, enthalpy, and the Joule–Thomson coefficient change with pressure and volume changes. In these cases, we can compute the particular thermodynamic function relative to a base temperature T^* and pressure P^* on a molar basis by changing one variable at a time. T^* and P^* are arbitrarily set at 0 °C and atmospheric pressure. On a molar basis

$$(\bar{h}_{P,T} - \bar{h}_{P^*,T^*}) = \int_{T^*,P^*}^{T,P^*} (\bar{C}_p) dT + \int_{P^*,T}^{P,T} \left[\bar{v} - T\left(\frac{\partial \bar{v}}{\partial T}\right)_P \right] dP \tag{2.26}$$

The bracketed term in the pressure correction is obtained from the Maxwell relations. The Maxwell relations were developed to interrelate various thermodynamic functions, such as enthalpy, with temperature and pressure.

For a system which is a solid at the base temperature and pressure, but a gas at the final temperature and pressure,

$$\left(\bar{h}_{P,T} - \bar{h}_{P*,T*}\right) = \int_{T*,P*}^{T_{\text{fusion}},P*} \left(\overline{C}_p\right)_{\text{solid}} dT + \Delta \overline{H}_{\text{fusion}} +$$

$$\int_{T_{\text{fusion}},P*}^{T_{\text{vaporization}},P*} \left(\overline{C}_P\right)_{\text{liquid}} dT + \Delta \overline{H}_{\text{vaporization}} +$$

$$\int_{T_{\text{vaporization}},P*}^{T,P*} \left(\overline{C}_P\right)_{\text{gas}} dT + \int_{P*,T}^{P,T} \left[\overline{v} - T\left(\frac{\partial \overline{v}}{\partial T}\right)_P\right] dP \quad (2.27)$$

If the compound is a liquid at $T*$ and $P*$, then the integral using $\left(\overline{C}_p\right)_{\text{solid}}$ and $\Delta \overline{H}_{\text{fusion}}$ are eliminated. The enthalpy change must be determined either from PVT data or an equation of state because of the need for $\overline{v}$ and $T(\partial \overline{v}/\partial T)_P$. Pressure–enthalpy data are given in charts (Figures 2.5 and 2.6) for Freon-12 and for perfect and real gases.

The reduced state approach is used to correct enthalpy for pressure changes. This correction is made by, first, computing the enthalpy changes for temperature or phase changes to bring a substance to the gaseous state, then calculating the enthalpy correction due to pressure for a gas, using charts (Figure 2.6).

Examples 2.1, 2.2, and 2.3 following illustrate a number of principles and con-

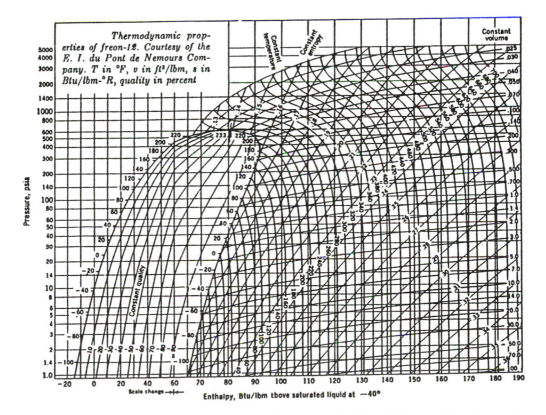

FIGURE 2.5 An enthalpy–pressure diagram such as this one for Freon-12 is a useful reference for calculating the energy of changes of state for real gases. (Reproduced with permission from DuPont.)

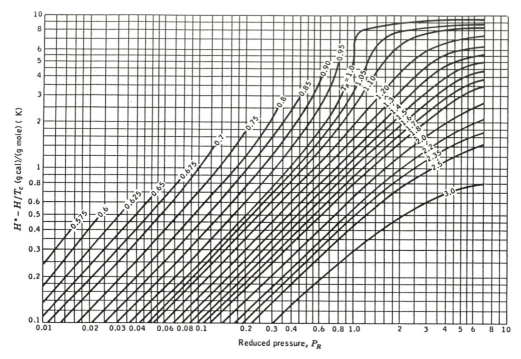

FIGURE 2.6 Although enthalpies for ideal gases do not change with changes in pressure, they do for real gases. This chart provides corrections for enthalpy correlated with reduced temperature and pressures. T_R is the reduced temperature (T/T_c), P_R is the reduced pressure (P/P_c), H^* is the ideal gas enthalpy, and H is the real gas enthalpy. Ordinate units are Btu/lb-mol-°R. (Reproduced with permission from reference 18. Copyright 1947 Wiley.)

cepts. Example 2.1 shows the path dependency of work, the point function nature of internal energy, and the simplification made possible by an ideal gas (i.e., internal energy depends only on temperature.) The second example illustrates the concept of the corresponding states principle, in dealing with the complex behavior of a multicomponent system of real gases. Example 2.3, which also applies the corresponding state concept, enables the practitioner to handle the combined effect of pressure and temperature on the enthalpy of a real gas. Examples 2.2 and 2.3 both emphasize how real systems can be handled with engineering thermodynamics.

EXAMPLE 2.1. WORK, INTERNAL ENERGY, AND AN IDEAL GAS

One lb-mass of nitrogen at 80 °F and 20 psia is contained in a vertical steel cylinder by a frictionless piston weighing 5.3 lb per in.² of surface. The whole apparatus is initially at equilibrium with its surroundings (air at 14.7 psia and 80 °F). The apparatus passes through the following cycle:

1. Apparatus is immersed in an ice bath and allowed to reach equilibrium.
2. Volume is reduced to half at the ice bath temperature.
3. Apparatus is removed and the piston is held in place by stops. The apparatus then comes to equilibrium with the surroundings.

Calculate:

1. Work appearing in the surroundings for each step and overall.
2. Internal energy change for each step and overall,

given:

C_p for nitrogen is 7 Btu/lb-mole-°R and
C_p for piston and cylinder are negligible.

The apparatus described is a control mass or closed system. The first law expression that governs it is

$$\Delta E = Q - W$$

However, since there are no kinetic or potential energy effects,

$$E = U$$

and

$$\Delta U = Q - W$$

Step 1 occurs at constant pressure

$$80 \text{ °F, 20 psia} \xrightarrow{\text{constant } P} 32 \text{ °F, 20 psia}$$

and under ideal gas conditions, that is, $P < 10$ atmospheres < 147 psia. For an ideal gas, internal energy is a function only of temperature and, from eq 2.17,

$$\Delta U_1 = C_v(\Delta T_1)n_{N_2}$$

The C_v for nitrogen is (again for ideal gas behavior)

$$C_p - C_v = R$$
$$C_v = C_p - R$$
$$C_v = (7 - 1.986) \text{ Btu/lb-mol-°R}$$
$$C_v = 5.014 \text{ Btu/lb-mol-°R}$$

and

$$\Delta U_1 = 5.014 \text{ Btu/lb-mol-°R } [(460 + 32) - (460 + 80) \text{ °R}] (1/28 \text{ lb-mol})]$$
$$\Delta U_1 = -8.61 \text{ Btu}$$

The work W_1 is given by

$$W_1 = P_{\text{surroundings}} \, \Delta \bar{v}$$

where $\bar{v}$ is the molar volume. The pressure, 20 psia, combines the weight of the piston (5.3 psia) and the atmospheric pressure (14.7 psia).

$$W_1 = (20 \text{ lb/in.}^2)(144 \text{ in.}^2/\text{ft}^2)(1/28 \text{ lb-mol})(\bar{v}_2 - \bar{v}_1)$$
$$W_1 = (20 \text{ lb/in.}^2)(144 \text{ in.}^2/\text{ft}^2)(1/28 \text{ lb-mol})(RT_2/P_2 - RT_1/P_1)$$

where R is 1544 ft-lbf/lb-mol/°R, T_1 is 540 °R, T_2 is 492 °R, and P_2 is P_1, which is 20 psia (144 in.^{2}/ft^2). Then,

$$W_1 = -2647 \text{ ft-lbf}$$

Now, consider the work done in step 2:

$$32 \text{ °F, } \bar{v}_2 \xrightarrow{\text{constant } T} 32 \text{ °F, } \bar{v}_2/2$$

Because we are at constant temperature with an ideal gas (20 psia < 147 psia) for which internal energy is a function only of temperature,

$$\Delta U_2 = 0$$

The work done

$$W_2 = n_{N_2} \int_{\bar{v}_2}^{\bar{v}_2/2} P d\bar{v} = n_{N_2} \int_{\bar{v}_2}^{\bar{v}_2/2} \frac{RT}{\bar{v}} d\bar{v}$$

$$W_2 = n_{N_2} [RT \ln \bar{v}]_{\bar{v}_2}^{\bar{v}_2/2} = n_{N_2} RT \ln 1/2$$

$$W_2 = (1/28 \text{ lb-mol})(1544 \text{ ft-lbf/lb-mol-°R})(492 \text{ °R}) \ln 1/2$$
$$W_2 = -18,800 \text{ ft-lbf}$$

Step 3 takes place at constant volume,

$$32 \text{ °F, } \bar{v}_2/2 \xrightarrow{\text{constant } v} 80 \text{ °F, } \bar{v}_2/2$$

Then, because we change only temperature for an ideal gas,

$$\Delta U_3 = n_{N_2} C_v[(80 + 460) \text{ °R} - (32 + 460) \text{ °R}]$$
$$\Delta U_3 = (1/28 \text{ lb-mol})(5.014 \text{ Btu/lb-mol/°R})(48 \text{ °R})$$
$$\Delta U_3 = 8.61 \text{ Btu}$$

Because $\bar{v}$ is constant, there is no work and

$$W_3 = 0$$

Summarizing

Step	ΔU (Btu)	W (ft-lbf)
1	−8.61	−2,647
2	0	−18,800
3	8.61	0
Overall	0	−21,447 (−27.6 Btu)

In SI units, the results would be

Step	ΔU (J)	W (J)
1	−9,100	−3,589
2	0	−25,492
3	9,100	0
Overall	0	−29,081

EXAMPLE 2.2. APPLYING THE THEORY OF CORRESPONDING STATES

What is the volume for an equal mass mixture of ethane and normal butane at a pressure of 5.065×10^6 N/m² and a temperature of 450 K? T_c and P_c values for ethane and n-butane, respectively, are 306 K, 425 K, 4.79×10^6 N/m² and 3.8×10^6 N/m².

To determine the thermodynamic properties of a gas mixture, we must know the quality or the mole fraction of each of the components in the mixture. Assume that we start with 1 kg of total mixture.

$$\text{kg-mol } C_2H_6 = 0.5 \text{ kg-mass}/30 \text{ kg-mass per kg-mol} = 0.01667$$

$$\text{kg-mol } n\text{-}C_4H_{10} = 0.5/58 = 0.00863$$

$$\text{mole fraction } C_2H_6 = 0.01667/0.02530 = 0.66$$

$$\text{mole fraction } n\text{-}C_4H_{10} = 0.00863/0.02530 = 0.34$$

$$P_c' = 0.66(4.79 \ 10^6 \text{ N/m}^2) + 0.34(3.8 \ 10^6 \text{ N/m}^2)$$

$$P_c' = 4.45 \times 10^6 \text{ N/m}^2$$

$$T_c' = 0.66(306 \text{ K}) + 0.34(425 \text{ K})$$

$$T_c' = 346.3 \text{ K}$$

then,

$$P_r' = P/P_c' = 5.065 \times 10^6 \text{ N/m}^2/4.45 \times 10^6 \text{ N/m}^2 = 1.13$$

and

$$T_r' = T/T_c' = 450 \text{ K}/346.3 \text{ K} = 1.30$$

Applying P_r' and T_r' values to the generalized compressibility chart (Figure 2.4), Z equals 0.83. Substituting Z

$$PV = ZnRT$$

$$V = ZnRT/P$$

$$V = $$
$$(0.83)(0.02530 \text{ kg-mol})(8314.41 \text{ m}^3\text{N/m}^2/\text{kg-mol-K})(450 \text{ K})/5.065 \times 10^6 \text{ N/m}^2$$

$$V = 0.01551 \text{ m}^3 \text{ per kg-mass of mixture}$$

EXAMPLE 2.3. APPLYING THE THEORY OF CORRESPONDING STATES

What is the enthalpy change due to a pressure change from 3.34 10^6 N/m^2 to 18.12 10^6 N/m^2 for *n*-hexane at 610 K? We must first calculate the values of T_r, P_{r_1}, and P_{r_2}:

$$T_r = 610 \text{ K}/507.9 \text{ K} = 1.2$$

$$P_{r_1} = 3.34 \ 10^6 \text{ N/m}^2/3.02 \ 10^6 \text{ N/m}^2 = 1.1$$

and

$$P_{r_2} = 18.12 \ 10^6 \text{ N/m}^2/3.02 \ 10^6 \text{ N/m}^2 = 6.0$$

At $T_r = 1.2$, $P_{r_1} = 1.1$ (from Figure 2.6)

$$(\bar{h}^\circ_{T,P^*} - \bar{h}_{T,P_{r_1}})/T_c = 2.05 \text{ cal/g-mol K}$$

At $T_r = 1.2$, $P_{r_2} = 6.0$ (from Figure 2.6).

$$(\bar{h}^\circ_{T,P^*} - \bar{h}_{T,P_{r_2}})/T_c = 7.0 \text{ cal/g-mol K}$$

Then,

$$(\bar{h}^\circ_{T,P^*} - \bar{h}_{T,P_{r_1}})/T_c - (\bar{h}^\circ_{T,P^*} - \bar{h}_{T,P_{r_2}})/T_c = (\bar{h}_{T,P_{r_2}} - \bar{h}_{T,P_{r_1}})/T_c$$

$$(\bar{h}_{T,P_{r_2}} - \bar{h}_{T,P_{r_1}}) = (2.05 - 7.0)(507.9 \text{ K})(4.184 \text{ J/cal})(1000 \text{ g/kg})$$

$$(\bar{h}_{T,P_{r_2}} - \bar{h}_{T,P_{r_1}}) = -1.052 \times 10^7 \text{ J/kg-mol}$$

APPLYING THE FIRST LAW TO FLOWS OF BOTH MASS AND ENERGY

Many industrial processes are systems in which mass flows across boundaries. Such systems must be represented in energy calculations by control volumes, a given volume into and from which both mass and energy can flow, rather than by control mass, the fixed mass considered in closed systems. We need a different expression for the first law, which is

$$\dot{Q} - \dot{W}_s = \dot{m}_{out}\,(e + PV) - \dot{m}_{in}\,(e + PV) + (dE/dt)_{\text{control volume}} \qquad (2.28)$$

The dots indicate quantities that are time derivatives. In the above equation, $\dot{Q}$ is the rate of heat transfer per unit time, $\dot{W}_s$ is the work per unit time done by the surroundings on the system via a turning shaft, that is, pumps, compressors, or turbines, e is energy per unit mass for the flowing fluid and includes all forms of energy—internal energy, kinetic energy, potential energy, etc.—except for expansion and compression work, PV is the expansion and compression work per unit mass, dE/dt represents all energy that accumulates in the control volume per unit time and includes all forms of energy such as internal, potential, and kinetic, $\dot{m}$ is mass flow rate into or out of the control volume per unit time, and V represents the specific volume (volume per unit mass).

If the system is operating at steady state, there is no energy accumulation (dE/dt is zero) or mass accumulation ($\dot{m}_{in}$ equals $\dot{m}_{out}$) in the system. Under steady-state conditions,

$$\dot{Q} - \dot{W}_s = \dot{m}(u + \Delta PV + \Delta PE + \Delta KE) \qquad (2.29)$$

where u is internal energy per unit mass, and

$$\dot{Q} - \dot{W}_s = \dot{m}(\Delta h + \Delta(Zg/g_c) + \Delta(\bar{V}^2/2g_c) \qquad (2.30)$$

In this equation, which says that the energy in a flow system consists of the change in enthalpy, potential energy, and kinetic energy, h is the change of enthalpy per unit mass, Z is the difference between the height of the flowing streams, $\bar{V}$ is the average velocity of the flowing stream, g is the force of gravity, and g_c is the factor applied to convert the units of mass times acceleration to force. (For more about the conversion factor g_c, refer to the box on the next page.) This equation is basic for all flow systems.

The following examples illustrate how the first law flow equation is applied. The recommended approach is to start with the equation in its complete form, followed by analysis to determine which terms can be neglected; in an adiabatic system, for example, $\dot{Q}$ is zero. Your analysis will show which contributions to the energy balance are important even without a quantitative solution. Finally, by virtue of the sign on the shaft work, $\dot{W}_s$, the equation will tell if you obtain work (positive sign) or put work into the system (negative sign).

In these flow cases, use the concept of a control volume where both energy and mass cross the system boundaries. In the case of control mass, only energy crosses the system boundary.

The approaches to solutions for the following examples will reappear in the chapters on fluid flow, heat transfer, mass transfer, and chemical reaction.

The Term g_c

The term g_c is the universal constant applied to convert the units of mass times acceleration to force. The basis of the need for g_c is Newton's second law, which is sometimes understood to be

$$\text{force} = (\text{mass})(\text{acceleration})$$

Actually, the equation is more correctly expressed as

$$f = (1/g_c)ma$$

Units of g_c are (mass) (acceleration)/(force). The term g_c must be included for all calculations involving g in all systems of units, even SI (Système International) and CGS (centimeter-gram-second), where the numerical value of g_c is equal to unity. The English system of units presents a special problem because g_c is 32.2 lb-mass-ft/lbf-s². Examples that include more extensive use of g_c are found in Chapter 3.

EXAMPLE 2.4. ADIABATIC FLOW AND A TWO-PHASE MIXTURE

Freon-12 is throttled adiabatically from a saturated liquid at 90 °F to a final state at which the pressure is 12 psia. What is the final temperature and the mole fraction in the vapor and liquid of the Freon-12?

The first law flow equation is

$$\dot{Q} - \dot{W}_s + \dot{m}_{in}\,(h_1 + Z_1 g/g_c + \overline{V}_1{}^2/g_c) = \dot{m}_{out}\,(h_2 + Z_2 g/g_c + \overline{V}_2{}^2/g_c) + \\ (dE/dt)_{\text{control volume}}$$

Since there is no accumulation and the system is at a steady state,

$$(dE/dt)_{\text{control volume}} = 0$$

and

$$\dot{m}_{in} = \dot{m}_{out} = \dot{m}$$

Because the system is adiabatic,

$$\dot{Q} = 0$$

and the system does not have any positive or negative work performed,

$$\dot{W}_s = 0$$

Finally, because the system has no change in height,

$$Z_1 = Z_2$$

and $\overline{V}_1$ equals $\overline{V}_2$. Hence,

$$\dot{m}\,h_1 = \dot{m}\,h_2$$

and

$$h_1 = h_2$$

For the given inlet conditions, the enthalpy h_1 of saturated liquid Freon-12 at 90 °F is 28.17 Btu/lb-mass (the far left curve in Figure 2.5). Therefore, h_2 is also 28.17 Btu/lb-mass at 12 psia.

The final enthalpy for the mixture is the sum of the mole fractions of the saturated liquid and of the saturated vapor multiplied by the enthalpies of liquid and vapor at 12 psia.

$$28.17 \text{ Btu/lb-mass} = (1 - x)(h_{\text{sat. liquid, 12 psia}}) + x(h_{\text{sat. vapor, 12 psia}})$$

$$28.17 \text{ Btu/lb-mass} = (1 - x)(4.24 \text{ Btu/lb-mass}) + x(74.0 \text{ Btu/lb-mass})$$

$$x = 0.34 \text{ mole fraction Freon-12 in vapor}$$

The temperature is determined by using the thermodynamic data chart (Figure 2.5). The intersection of the horizontal 12 psia line with the left or right side of the dome gives a temperature of -32 °F for saturated liquid or saturated vapor, respectively.

EXAMPLE 2.5. ACCOUNTING FOR MECHANICAL WORK

A hydroelectric project in a developing country has a volumetric throughput of 1.2 m³/s. The water flowing in the river at atmospheric pressure and 20 °C falls vertically for 300 m and passes through a turbine. The water exits the turbine at atmospheric pressure and 20.5 °C. What is the power output of the turbine?

Once again the first law flow equation applies.

$$\dot{Q} - \dot{W}_s + \dot{m}_{\text{in}}\,(h_1 + Z_1 g/g_c + \overline{V}_1^2/g_c) =$$
$$\dot{m}_{\text{out}}\,(h_2 + Z_2 g/g_c + \overline{V}_2^2/g_c) + (dE/dt)_{\text{control volume}}$$

Because the system has a steady flow ($\dot{m}_{\text{in}} = \dot{m}_{\text{out}} = \dot{m}$), there is no change in velocity from inlet to outlet ($\overline{V}_1$ equals $\overline{V}_2$), and there is no accumulation of energy ($(dE/dt)_{\text{control volume}}$ equals zero). Also, the turbine may be considered adiabatic, so $\dot{Q}$ is zero. This leaves

$$-\dot{W}_s = \dot{m}[(h_2 - h_1) + (Z_2 - Z_1)g/g_c]$$

Calculating Δh from $C_p \Delta T$ results in

$$\dot{W}_s/\dot{m} = (4.1854 \times 10^3 \text{ J/kg-°C})(20.5 - 20)°C + (300 \text{ m})(9.81 \text{ J/kg-m}^2)$$
$$\dot{W}_s/\dot{m} = 5036 \text{ J/kg}$$
$$\dot{W}_s = (5036 \text{ J/kg})(1.2 \text{ m}^3/\text{s})(1 \times 10^3 \text{ kg/m}^3)$$
$$\dot{W}_s = 6.04 \times 10^6 \text{ J/s} = 6.04 \times 10^6 \text{ W} = 6.04 \text{ MW}$$

THE SECOND LAW OF THERMODYNAMICS— THE LIMITS OF EFFICIENCY

While useful, the first law does not limit the amount of heat transfer possible. If we were to calculate the energy transferred between two fluids in a heat exchanger, for example, we would realize that the temperature of the cooler fluid could not rise above that of the warmer fluid. But the first law, which allows us to calculate the heat transferred, does not restrict the maximum temperature rise of the cooler fluid. Hence, it is necessary to formulate another law which governs possibility. This second law is stated: "All spontaneous processes are irreversible."

Another version of the second law is: "It is impossible to construct a machine operating in a cycle which will produce no other effect than the transfer of heat from a cooler to a warmer source."

DEFINING ENGINE EFFICIENCY—THE CARNOT CYCLE

Related to the second law is the concept of the perfectly reversible cycle, the Carnot cycle, which theoretically offers the most heat transfer achievable in any process. The Carnot cycle is depicted in a *PV* diagram (lower graph in Figure 2.7) in a series of four steps:

1. A gas undergoes a reversible adiabatic compression with temperature rising from T_1 to T_2 (step d to a).
2. The gas expands isothermally and reversibly at T_2 until it reaches a new *PV* combination (step a to b).
3. The gas expands reversibly and adiabatically until the temperature drops to T_1 (step b to c).
4. The gas returns to the original state by undergoing reversible, isothermal compression at T_1 (step c to d).

Application of the Carnot cycle is not to limited to pressure-volume situations but can be applied to any work mode (i.e., electrochemical, magnetic, etc.).

No engine operating between two reservoirs of energy (i.e., hot and cold) can be more efficient than a Carnot engine for the same temperature and conditions. The Carnot engine's efficiency η is defined as

$$\eta = 1 - T_{\text{cold}}/T_{\text{hot}} \qquad (2.31)$$

where T_{cold} and T_{hot} represent the absolute temperatures of the reservoirs. The ratio of heat transferred is directly proportional to the ratio of the hot and cold reservoir temperatures,

$$|Q_{\text{hot}}|/|Q_{\text{cold}}| = T_{\text{hot}}/T_{\text{cold}} \qquad (2.32)$$

where $|Q_{\text{hot}}|$ and $|Q_{\text{cold}}|$ are the absolute amounts of heat transferred from the hot and to the cold reservoirs, respectively. Based on the above with sign convention,

$$Q_{\text{hot}}/T_{\text{hot}} = Q_{\text{cold}}/T_{\text{cold}} \qquad (2.33)$$

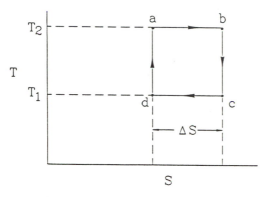

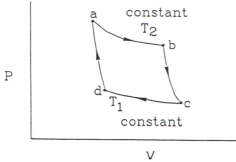

FIGURE 2.7 The Carnot cycle, depicted here on both a temperature–entropy and pressure–volume diagram, represents perfect engine efficiency in an expansion–compression work system.

We can apply the concept of the Carnot cycle to a generalized cyclic work situation (see Figure 2.8) by dividing the overall cycle into a series of small Carnot cycles. For this case,

$$Q_1/T_1 + Q_2/T_2 = 0$$

and

$$Q_3/T_3 + Q_4/T_4 = 0$$

so that

$$Q_1/T_1 + Q_2/T_2 + Q_3/T_3 + Q_4/T_4 = 0$$

Hence, for the generalized cyclic work situation,

$$\sum_i Q_i/T_i = 0 \qquad (2.34)$$

For a reversible case, then,

$$\oint_{\text{reversible}} d'Q/T = 0 \qquad (2.35)$$

and we define a new thermodynamic property, entropy, S, as

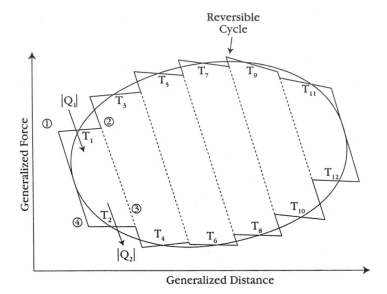

FIGURE 2.8 Dividing a complex pressure–volume work cycle into a series of small Carnot cycles enables us to predict the maximum work possible from the system. For a reversible cycle such as the one shown here, the change in entropy is zero. (Adapted from reference 19.)

$$d'Q_{\text{reversible}}/T = dS$$

The entropy equations for different systems are presented in Table 2.3.

ENTROPY

The entropy change of a system indicates its possibility as a process. In general, the entropy change of a process must be either zero (the process is reversible) or positive (the process is irreversible). If the entropy change is negative, then the process cannot occur.

TABLE 2.3 Forms of the Entropy Expression

System	First Law	Second Law[a]
Isolated (no exchange of energy or mass with surroundings)	$dE = 0$	$dS \geq 0$
Closed, control mass (exchanges only energy with surroundings)	$dE = d'Q - d'W$	$dS \geq d'Q/T$
Open, control volume (exchanges both energy and mass with surroundings)	$(dE/dt)_{\text{control volume}} = \dot{Q} - W_s + \dot{m}_{\text{in}}\,(e + PV) - \dot{m}_{\text{out}}(e + PV)$	$dS/dt \geq \sum \dot{Q}/T + \sum_{\text{in}} s\dot{m} - \sum_{\text{out}} s\dot{m}$

[a]$s\dot{m}$ refers to entropy per mass.

The energy used by mechanical devices such as pumps, turbines, and compressors illustrates the entropy concept. If we calculate the entropy of a device in terms of its process reversibility, we're calculating the maximum possible output of a pump or turbine, and the maximum input needed for a compressor. In real life, however, the irreversibility of operating a piece of equipment reduces the work output of turbines and pumps and increases the work needed from a compressor.

The entropy for a gaseous system is calculated at a given pressure and temperature by working from a standard, known base temperature and pressure

$$(\bar{S}_{P,T} - \bar{S}_{P^*,T^*}) = \int_{T^*,P^*}^{T,P^*} (\bar{C}_p)_{gas}/T \, dT - \int_{P^*,T}^{P,T} (\partial \bar{V}/\partial T)_P \, dP \qquad (2.36)$$

The entropy of a system that starts as a solid can be calculated in steps, working from solid to liquid to gas

$$(\bar{S}_{P,T} - \bar{S}_{P^*,T^*}) = \int_{T^*,P^*}^{T_{fusion},P^*} (\bar{C}_p)_{solid}/T \, dT + \Delta \bar{H}_{fusion}/T_{fusion} +$$

$$\int_{T_{fusion},P^*}^{T_{vaporization},P^*} (\bar{C}_p)_{liquid}/T \, dT + \Delta \bar{H}_{vaporization}/T_{vaporization} +$$

$$\int_{T_{vaporization}P^*}^{T,P^*} (\bar{C}_p)_{gas}/T \, dT -$$

$$\int_{P^*,T}^{P,T} (\partial \bar{V}/\partial T)_P \, dP \qquad (2.37)$$

If the system is a liquid at T^* and P^*, then the integral for $(\bar{C}_p)_{solid}$ and the $\Delta H_{fusion}/T_{fusion}$ terms are eliminated. Refer either to PVT data or an equation of state to evaluate the entropy change with pressure. Because

$$(\partial \bar{V}/\partial T)_P = R/P \qquad (2.38)$$

for an ideal gas, the last integral in eq 37, is the following for that system:

$$\int_{P^*,T}^{P,T} (\partial \bar{V}/\partial T)_P \, dP = \int_{P^*,T}^{P,T} R/P \, dP \qquad (2.39)$$

A combined chart of enthalpy and entropy for water is shown in Figure 2.9. When the pressure exceeds that for an ideal gas (greater than 10 atmospheres), use the reduced state concept and entropy values from a chart (Figure 2.10).

As a thermodynamic property, entropy gives us an understanding of the interaction between energy and matter, as we'll see by working Examples 2.6 and 2.7. In Example 2.6, we'll first use the behavior of entropy to derive the cycle for a refrigeration unit on a temperature-entropy schematic. Then we'll apply a combination of entropy and enthalpy behavior to find all of the pertinent values needed to calculate the refrigerator's efficiency, the Coefficient of Performance.

Example 2.7 illustrates the concept of irreversibility with a calculation of a turbine's mechanical efficiency. Without the use of entropy, we would not be able to delineate the real work obtained from the turbine.

EXAMPLE 2.6. EVALUATING THE PERFORMANCE OF A REFRIGERATOR

A compression engine uses Freon-12 to maintain a cold room. The expansion/compression cycle, with the temperature and pressures noted, is depicted in Figure 2.11. In the cycle, the gas is compressed isentropically to 99 psia (path AB), condensed at constant pressure (BC), expanded to 24 psia (CD), and, finally, vaporized at constant temperature (DA). Evaluate the refrigerator's performance.

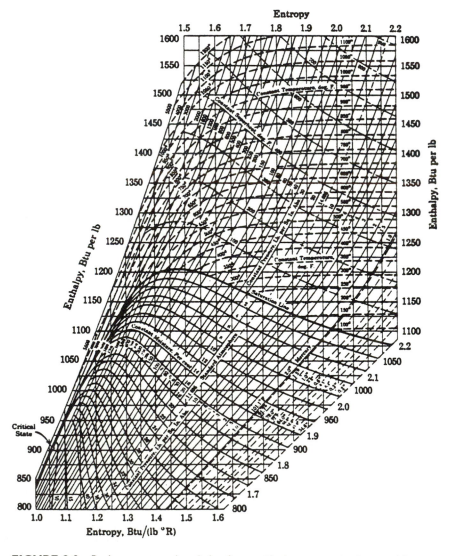

FIGURE 2.9 Both entropy and enthalpy for nonideal compounds change with pressure and temperature, as shown in this chart for water. (Reproduced with permission from reference 4. Copyright 1955 University of Wisconsin.)

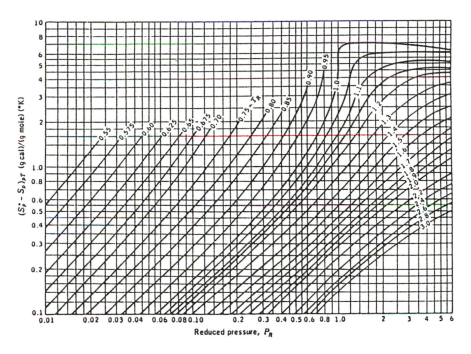

FIGURE 2.10   The correction needed for entropy for pressure in a system above the ideal gas state (>10 atm), can be found from this chart if the reduced pressures and temperatures are known. T_R is the reduced temperature (T/T_c), P_R is the reduced pressure (P/P_c), S_P^* is the entropy of an ideal gas, and S_P is the entropy of a real gas. The ordinate units are Btu/lb-mol-°R. (Reproduced with permission from reference 4. Copyright 1955 University of Wisconsin.)

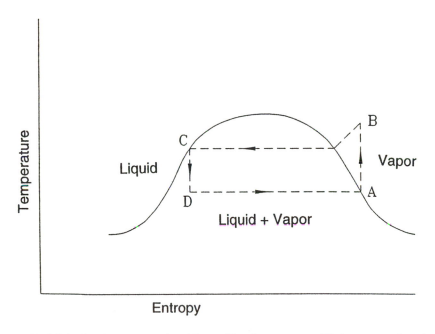

FIGURE 2.11   Here is a cycle of Freon-12 refrigerant on a TS diagram as described in Example 2.6. Point A is saturated vapor at 0 °F. Point B is 99 psia. Point C is saturated liquid at 99 psia. Point D is 24 psia.

To evaluate the efficiency of a refrigerator, calculate the coefficient of performance (COP) for the cycle, defined as

$$COP = (h_A - h_D)/[(h_B - h_C) - (h_A - h_D)]$$

Starting with point A, (Figure 2.5) the enthalpy h_A for the saturated vapor at 0 °F is 78.2 Btu/lb-mass, and the entropy S_A is 0.171 Btu/lb-mass-°R.

Path AB is at constant entropy. Hence

$$S_A = S_B = 0.171 \text{ Btu/lb-mass-°R}$$

and for this entropy and 99 psia, the enthalpy h_B (from Figure 2.6) is 89.3 Btu/lb-mass.

The value of h_C is 26.3 Btu/lb-mass for a saturated liquid at 99 psia. The entropy S_C is 0.055 Btu/lb-mass °R (from Figure 2.5).

The path from C to D is at constant entropy. Hence,

$$S_D = S_C = 0.055 \text{ Btu/lb-mass-°R}$$

Point D is in the two-phase region, and so, we make use of the quality x in the equations for entropy and enthalpy. Thus,

$$S_D = S_A - x(\Delta H_{\text{vaporization},0°F}/460 \text{ °R})$$

$$h_D = h_A - x(\Delta H_{\text{vaporization}, 0 \text{ °F}})$$

Eliminating x and solving for h_D gives h_D equal to 24.9 Btu/lb-mass.

Finally, calculating the COP

$$COP = (78.2 - 24.9)/[(89.3 - 26.3) - (78.2 - 24.9)]$$

$$COP = 5.5$$

The larger the value of the COP, the better. The value from the Carnot cycle is the limit.

EXAMPLE 2.7.   EFFICIENCY LOSSES OF A TURBINE

A turbine operated with steam has an energy output of 4.51×10^5 J/kg. The operating conditions are

Entrance	Exit
$\overline{V}_1 = 76.3$ m/s	$\overline{V}_2 = 152.2$ m/s
$P_1 = 1.52 \times 10^6$ N/m²	$P_2 = 1.52 \times 10^5$ N/m²
$T_1 = 393$ °C	

Calculate the turbine's mechanical efficiency.

The maximum work output will take place in an adiabatic reversible process. For this process with steady flow (eq 30)

$$\dot{Q} - \dot{W}_s = \dot{m}\,(\Delta h + \Delta Z g/g_c + \Delta \overline{V}^2/2g_c)$$

For the turbine, there is no change in potential energy nor transfer of heat (adiabatic). Therefore,

$$-\dot{W}_s/\dot{m} = \Delta h + \Delta \bar{V}^2/2g_c$$

From Figure 2.9, h_1 for the system at T_1 of 393 °C and P_1 of 1.52×10^6 N/m^2 is 3.24×10^6 J/kg. Likewise S_1 (at the same conditions) is 7.22×10^3 J/kg K.

To get h_2, follow an isentropic path (i.e., $S_2 = S_1$) to get to $P_2 = 1.52 \times 1.0^5$ N/m^2 (again on Figure 2.9). Therefore, S_2 equals 7.22×10^3 J/kg-K. For these values of S_2 and P_2,

$$T_2 = 116 \text{ °C}$$

$$h_2 = 2.7 \times 10^6 \text{ J/kg}$$

$$-\dot{W}_s/\dot{m} = (h_2 - h_1) + (\bar{V}_2{}^2 - \bar{V}_1{}^2)/2g_c$$

$$-\dot{W}_s/\dot{m} = [(2.7 - 3.24) \times 10^6 \text{ J/kg}] + [(152.2)^2 - (76.3)^2 \text{ m}^2/\text{sec}^2/2]$$

and

$$-\dot{W}_s/\dot{m} = -5.31 \times 10^5 \text{ J/kg}$$

The turbine's efficiency is calculated by the comparing the observed output to the theoretical output:

$$\text{mechanical efficiency} = (4.51 \times 10^5 \text{ J/kg})(100)/5.31 \times 10^5 \text{ J/kg} = 85\%$$

The difference between the actual and calculated work output is due to irreversibilities.

GIBBS FREE ENERGY, CHEMICAL POTENTIAL, FUGACITY, AND ACTIVITY

As pointed out earlier, we must describe the character of phase equilibrium to design, analyze, or optimize separation processes. It's always better to work from a base of actual data. When they aren't available, we need methods for estimating phase equilibrium data.

Equilibrium data are calculated by using a number of interrelated thermodynamic properties. The Gibbs free energy F is the starting point because it is the basis for equilibrium. For a mole, F is defined as

$$F = H - TS = U + PV - TS \tag{2.40}$$

and

$$dF = dU + PdV + VdP - TdS - SdT \tag{2.41}$$

From a combination of the first and second laws, we know that

$$dU = TdS - PdV \tag{2.42}$$

and therefore,

$$dF = VdP - SdT \tag{2.43}$$

When the temperature is constant,

$$dF = VdP \tag{2.44}$$

For one mole of ideal gas,

$$dF = (RT/P)dP = RT\, d(\ln P) \tag{2.45}$$

Generally, for a pure substance,

$$dF = RT\, d(\ln f) \tag{2.46}$$

where f is the fugacity, a thermodynamic function. Fugacity is described qualitatively as the "escaping tendency" of a component from a solution into a vapor, a type of partial pressure. Fugacity equals the partial pressure only for a vapor that behaves like an ideal gas. If equal in both phases, fugacity is a function that signifies the existence of equilibrium. Next, integrating eq 46 from a standard state (designated by the zero superscript),

$$F - F^0 = RT \ln f/f^0 = \int_{P^0}^{P} VdP \tag{2.47}$$

As the pressure approaches zero,

$$\lim_{P \to 0} f/P = 1 \tag{2.48}$$

For a solution or components, we introduce the concept of chemical potential μ_i related to fugacity. For a given component i

$$d\mu_i = RT\, d \ln \bar{f_i} \tag{2.49}$$

where the bar indicates the fugacity of a component in solution. Chemical potential is also related to the partial differentials of the thermodynamic functions of internal energy, enthalpy, entropy, Gibbs free energy, and Helmholtz free energy A by

$$\mu_i = (\partial U/\partial n_i)_{S,V,n_j} = (\partial H/\partial n_i)_{S,P,n_j} = (\partial F/\partial n_i)_{S,T,n_j} = (\partial A/\partial n_i)_{T,V,n_j} \tag{2.50}$$

where i is not equal to j, n_i represents moles of i, and n_j the individual moles of other components. Note that V, H, F, and A are extensive properties. The moles of all components except i are held constant.

J. Willard Gibbs used the chemical potential to develop his treatment of equilibrium. When the chemical potentials of a component in various phases are the same, the system is in equilibrium. The connection between chemical potential and fugacity makes it possible to use fugacity to determine whether equilibrium exists. Fugacity has some real advantages in evaluating equilibrium because it does not vary over as wide a numerical range as chemical potential.

The final thermodynamic property of phase equilibrium is the activity a_i which is related to both chemical potential and fugacity by

$$\mu_i - \mu_i^0 = RT \ln \bar{f}_i/f_i^0 = RT \ln a_i \qquad (2.51)$$

At equilibrium

$$\mu_i^{(1)} = \mu_i^{(2)} = \mu_i^{(3)} = \cdots \mu_i^{(n)} \qquad (2.52)$$

where the superscripts indicate different phases. For a vapor–liquid system at equilibrium,

$$\mu_i^L = \mu_i^V \qquad (2.53)$$

and

$$RT \ln \bar{f}_i^L/f_i^0 + \mu_i^0 = RT \ln \bar{f}_i^V/f_i^0 + \mu_i^0 \qquad (2.54)$$

so that

$$\bar{f}_i^L = \bar{f}_i^V \qquad (2.55)$$

As shown earlier,

$$\mu_i = (\partial F/\partial n_i)_{S,T,n_j} = \bar{F}_i$$

and

$$(\partial \mu_i/\partial P)_T = (\partial \bar{F}_i/\partial P)_T = \bar{v}_i$$

where the bar indicates the partial molal volume for component i, so that

$$RT(\partial \ln \bar{f}_i/\partial P)_T = \bar{v}_i \qquad (2.56)$$

and

$$RT \ln \bar{f}_i/f_i^0 = \int_{P^0}^{P} \bar{v}_i \, dP \qquad (2.57)$$

Then,

$$RT \ln (y_i P/y_i P^0) = \int_{P^0}^{P} RT/P \, dP \qquad (2.58)$$

where y_i is the mole fraction. Subtracting eq 58 from eq 57 yields

$$RT \ln (\bar{f}_i/y_i P) - RT \ln (f_i^0/y_i P^0) = \int_{P^0}^{P} (\bar{v}_i - RT/P) dP \qquad (2.59)$$

If P^0 approaches zero,

$$RT \ln (\bar{f}_i/y_i P) = \int_{0}^{P} (\bar{v}_i - RT/P) dP \qquad (2.60)$$

and, for a pure component i,

$$RT \ln f_i = RT \ln P + \int_0^P (\bar{v} - RT/P) dP \tag{2.61}$$

Subtracting eq 61 from eq 60 gives

$$RT \ln (\bar{f}_i/y_i f_i) = \int_0^P (\bar{v}_i - \bar{v}) dP \tag{2.62}$$

Assuming the solution is ideal, that is, the solution volume equals the sum of the partial volumes, and the pressure is low (Amagat's law holds), then,

$$\bar{v}_i - \bar{v} = 0$$

so that

$$\ln (\bar{f}_i/y_i f_i) = 0 \tag{2.63}$$

and

$$\bar{f}_i = y_i f_i \tag{2.64}$$

This is known as the Lewis and Randall rule.

APPLIED PHASE EQUILIBRIUM AND THE DISTRIBUTION COEFFICIENT

The distribution of a given component between phases controls the movement of energy (and mass) across the phase boundary. The distribution is governed by the thermodynamic functions discussed in the previous section. However, to obtain such distribution data, various assumptions may have to be made about the system, such as assuming an ideal solution, ideal gas, etc. Furthermore, generalized techniques are necessary when experimental data is lacking.

This section discusses various methods and techniques for obtaining information about the distribution of components among phases at equilibrium.

The distribution coefficient for component i is

$$K_i = (y_i/\bar{f}_i^V)/(x_i/\bar{f}_i^L) \tag{2.65}$$

where x_i and y_i are liquid and vapor mole fractions, and $\bar{f}_i^V$ and $\bar{f}_i^L$ are the vapor and liquid phase fugacities of component i. At equilibrium,

$$\bar{f}_i^V = \bar{f}_i^L \tag{2.66}$$

and

$$K_i = y_i/x_i \tag{2.67}$$

Also,

$$K_i = f_i^L/f_i^V \tag{2.68}$$

because

$$\bar{f}_i^V / y_i = f_i^V$$

and

$$\bar{f}_i^L / x_i = f_i^L$$

(of course, with Amagat's law in effect).

For a pure substance,

$$\ln f_i^L / f_{i,P^+}^L = 1/RT \int_{P^+}^{P} v_i^L dP \qquad (2.69)$$

where P^+ is the vapor pressure at T. At equilibrium,

$$f_{i,P^+}^L = f_{i,P^+}^V \qquad (2.70)$$

so that

$$f_i^L = f_{i,P^+}^V \exp\left[v_i^L (P - P^+)/RT\right] \qquad (2.71)$$

and

$$K_i = (f_{i,P^+}^V / f_i^V)\exp\left[v_i^L(P - P^+)/RT\right] \qquad (2.72)$$

Equation 72 assumes Amagat's law and a metastable equilibrium. The v_i^L is the saturated liquid molar volume of i at T.

When the vapor is an ideal gas,

$$K_i = (P^+/P) \exp[v_i^L(P - P^+)/RT] \qquad (2.73)$$

and if P is almost equal to P^+, then,

$$K_i = P^+/P \qquad (2.74)$$

which is Raoult's law.

For nonideal solutions (i.e., Amagat's law does not hold),

$$\bar{f}_i^V = \gamma_i^V y_i f_i^V \qquad (2.75)$$

and

$$\bar{f}_i^L = \gamma_i^L x_i f_i^L \qquad (2.76)$$

where γ is an *activity coefficient*. These equations yield

$$K_i = (\gamma_i^L f_{i,P^+}^V / \gamma_i^V f_i^V)\exp\left[v_i^L (P - P^+)/RT\right] \qquad (2.77)$$

The various forms of K_i are summarized in Table 2.4.

TABLE 2.4 Forms of the Distribution Coefficient K_i

K_i	Type of Solution	Special Conditions
$f_{i,P^+}^V/f_i^V \exp [v_i^L(P - P^+)/RT]$	Ideal liquid and vapor solution (Amagat's law)	Metastable equilibrium
$f_{i,P^+}^V/f_i^V$	Ideal gas mixture	No metastable equilibrium
$(P^+/P) \exp[v_i^L (P - P^+)/RT]$	Ideal liquid solution (Amagat's law)	Ideal gas vapor
P^+/P	Ideal liquid solution (Amagat's law)	$P \cong P^+$ ideal gas vapor
$(\gamma_i^L f_{i,P^+}^V/\gamma_i^V f_i^V) \exp [v_i^L (P - P^+)/RT]$	Nonideal vapor and liquid solutions (Amagat's law does not hold)	None
$(\gamma_i^L f_{i,P^+}^V/f_i^V) \exp [v_i^L(P - P^+)/RT]$	Nonideal liquid solution (Amagat's law does not hold)	Ideal gas vapor

Evaluation of the distribution coefficient requires values of fugacities for pure substances. These fugacities can be determined by a number of methods using the form

$$\ln f/P = -1/RT \int_0^P (RT - v)dP \tag{2.78}$$

and one of the following: (1) assuming an ideal gas, (2) using an equation of state (The result of applying the generalized theory of corresponding states to the equation of state to get v and then calculating the fugacities is shown in Figures 2.12 and 2.13.), and (3) calculating the fugacities directly from PVT data.

Distribution coefficients can, also, be calculated directly from other equations of state. Several tabulations of such results are based on the Benedict–Webb–Rubin equation of state (6):

$$P = B_0 RT\bar{d}^2 + bRT\bar{d}^3 - A_0\bar{d}^2 - a\bar{d}^3 + \alpha a\bar{d}^6 - c_0\bar{d}^2/T^2 +$$

$$(c\bar{d}^3/T^2)(1 + \gamma\bar{d}^2)e^{-\gamma\bar{d}^2} + RT\bar{d} \tag{2.79}$$

The tabulations were done by calculating the liquid and vapor fugacity. The first tabulation of distribution coefficients resulted in 324 Kellogg equilibrium charts (7) of twelve hydrocarbons, liquid and vapor, as functions of pressure, temperature, and molal average boiling points (MABP):

$$\text{MABP}_{\text{liquid}} = \sum_i x_i B_i \tag{2.80}$$

and

$$\text{MABP}_{\text{vapor}} = \sum_i y_i B_i \tag{2.81}$$

where B_i is the boiling point of component i.

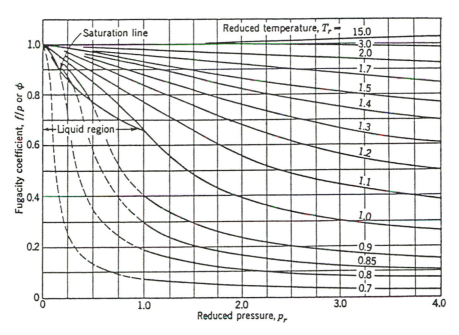

FIGURE 2.12 At high pressures, the partial vapor pressure of a nonideal gas can be found from the fugacity. This chart shows the generalized relationship of fugacity to reduced pressure and temperature based on the corresponding states theory. (Reproduced with permission from reference 4. Copyright 1955 University of Wisconsin.)

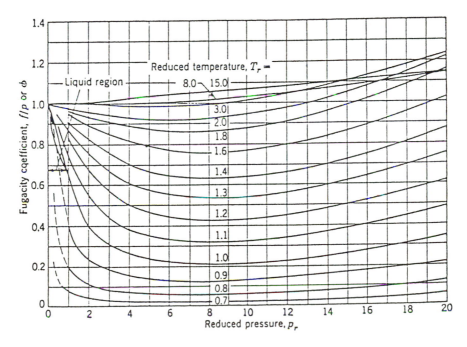

FIGURE 2.13 At high pressures, the fugacity can be found from this chart based on reduced pressure and temperature. (Reproduced with permission from reference 4. Copyright 1955 University of Wisconsin.)

These calculations involved eight to ten steps for each component. Later, the 324 charts required for Kellogg's calculations were reduced to a smaller number by De Priester (8). De Priester, also, developed a set of nomographs of equilibrium constants (Figures 2.14 and 2.15) for pressures less than 800 psia, to determine K_is.

Edmister and Ruby (9) refined the earlier treatments to reflect the reduced temperature and pressure (the system temperature in absolute units divided by the component critical temperature, and the system pressure divided by the component critical pressure) as well as MABP.

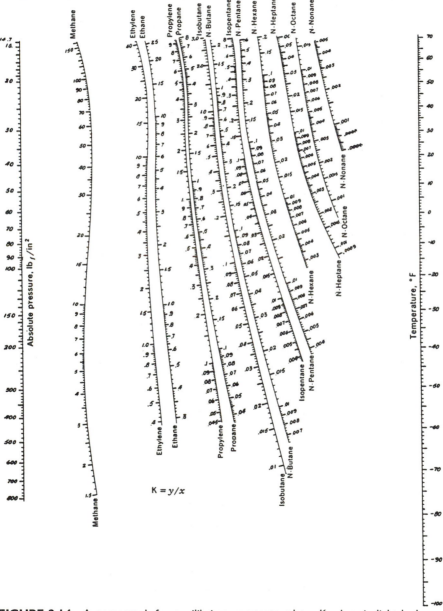

FIGURE 2.14 A nomograph for equilibrium constants relates K values in light hydrocarbon systems to temperature and pressure. This nomograph is for low temperatures. To find the value of K, draw a straight line between the known temperature and pressure and read the K value for each compound. (Reproduced with permission from reference 8. Copyright 1953 AIChE.)

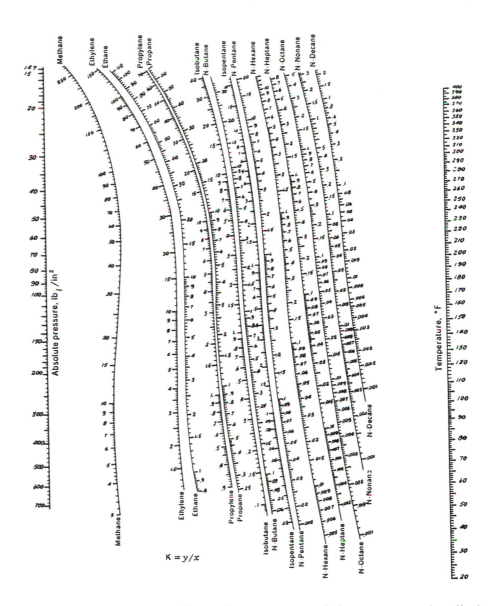

FIGURE 2.15 A nomograph for equilibrium constants at high temperatures relates K values in light hydrocarbon systems to temperature and pressure. (Reproduced with permission from reference 8. Copyright 1953 AIChE.)

An alternate method of calculating the distribution coefficient is based on converging pressure value. This pressure is the pressure at which distribution coefficient values appear to converge to unity on a plot of $\ln K_i$ versus $\ln P$. These data are presented in chart form (*10*) or nomographs (*11*).

Example 2.8 illustrates how the different techniques can be used to calculate distribution coefficients. The calculated values of the coefficients will, then, be compared to those from actual data to highlight the advantages and disadvantages of different approaches. Often no actual data exists, so that the practitioner must rely on calculated values when designing mass transfer processes such as distillation and absorption.

EXAMPLE 2.8.   CALCULATING DISTRIBUTION COEFFICIENTS

Calculate distribution coefficients at 37.8 °C and 6.078×10^6 N/m^2 for a system consisting of methane, ethylene, and isobutane for the following cases: (1) Raoult's law applies, (2) ideal gas mixture, no metastable equilibrium, (3) ideal liquid and vapor solutions, metastable equilibrium, and (4) using the data from Figures 2.14, 2.15.

Solution 1. If Raoult's law applies, then,

$$K_i = P^+/P$$

Vapor pressures for the compounds at 37.8 °C are as follows: methane (3.58×10^7 N/m^2), ethylene (8.33×10^6 N/m^2), and isobutane (5.02×10^5 N/m^2). Hence,

$$K \text{ for methane} = 3.58 \times 10^7 \text{ N/m}^2/6.078 \times 10^6 \text{ N/m}^2 = 5.9$$

$$K \text{ for ethylene} = 8.33 \times 10^6 \text{ N/m}^2/6.078 \times 10^6 \text{ N/m}^2 = 1.37$$

and

$$K \text{ for isobutane} = 5.02 \times 10^5 \text{ N/m}^2/6.078 \times 10^6 \text{ N/m}^2 = 0.0825$$

Solution 2. Ideal gas solution, no metastable equilibrium.
    Solving this case requires the use of Figures 2.12 and 2.13. To use these charts, we must calculate T_r, P_r, and P_r^+ (P^+/P_{critical}) for all three compounds.

Compound	T_c (K)	T/T_c	P_c (N/m^2)	P/P_c	P^+/P_c
Methane	190.7	1.63	4.66×10^6	1.30	7.68
Ethylene	283.1	1.10	5.20×10^6	1.17	1.61
Isobutane	409	0.760	3.67×10^6	1.66	0.137

With the data, we can find f_{i,P^+}^V and f_i^V from the figures. Hence,

Compound	$f_{i,P^+}^V/P^+$	f_{i,P^+}^V (N/m^2)	f_i^V/P	f_i^V (N/m^2)	K
Methane	0.785	2.81×10^7	0.930	5.65×10^6	4.97
Ethylene	0.66	5.5×10^6	0.720	4.38×10^6	1.26
Isobutane	0.90	4.52×10^5	0.096	5.83×10^5	0.77

Solution 3. Ideal liquid and vapor solutions, metastable equilibrium

$$K = f_{i,P^+}^V / f_i^V \exp \left[v_i^L (P - P^+)/RT \right]$$

The liquid v_i^L s at 37.8 °C (extrapolated for methane and ethylene) are for methane (0.0745 m³/kg-mol), for ethylene (0.1362 m³/kg-mol), and for isobutane 0.108 m³/kg-mol). Note that the calculation involves multiplying the Ks of Solution 2 by the exponential term.

Compound	exp $[v_i^L(P - P^+)/RT]$	K
Methane	0.000198	9.8×10^{-4}
Ethylene	0.88870	1.12
Isobutane	1.5605	1.21

Solution 4. Use of Figures 2.14 and 2.15.

Compound	K
Methane	3.2
Ethylene	1.15
Isobutane	0.165

An overall comparison of results from the different techniques is given below. Case 5 represents experimental data.

Compound	Case 1	Case 2	Case 3	Case 4	Case 5
Methane	5.90	4.97	9.8×10^{-4}	3.20	3.06
Ethylene	1.37	1.26	1.12	1.15	1.25
Isobutane	0.0825	0.770	1.21	0.165	0.284

Note the extreme deviation of the Case 3 K value for methane. This occurs because the 37.8 °C temperature is far above methane's critical point.

ACTIVITY COEFFICIENTS

The forms of the distribution coefficients in Table 2.4 require values for gas and liquid phase activity coefficients. In some instances, gas phase activities are close to unity. This is true for ideal gas behavior. A general form for the vapor activity coefficient is

$$RT \ln \gamma_i^V = \int_0^P (\bar{v}_M^V - \bar{v}_i^V) dP + \int_0^P \partial \bar{v}_M^V / \partial y_i dP - \sum_j Y_j \int_0^P \partial \bar{v}_M^V / \partial y_j \, dP \qquad (2.82)$$

where $\bar{v}_M^V$ is the molar volume of the vapor mixture and the ys are the vapor mole fractions. The best way to use eq 82 is to insert actual PVT data. When actual data are not available, equations of state giving the best fit can be used to find $\bar{v}_M^V$ as a function of composition and pressure at T for substitution in eq 82.

For example, the corresponding state theory can be substituted for $\bar{v}_M^V$, which gives

$$\ln \gamma_i^V + \ln f_i/P = \ln f_{\text{mix}}/P + (\bar{h}^\circ - \bar{h})/RTT_c' + ((Z-1)/P_c')(P_c' - P) \qquad (2.83)$$

The T_c' ($T_c' = \Sigma y_i T_{c_i}$) is the pseudocritical temperature of the mixture, P_c' ($P_c' = \Sigma y_i P_{c_i}$) the pseudocritical pressure, and $(\bar{h}^\circ - \bar{h})$ the enthalapy correction for pressure (from Figure 2.6). With ideal gas behavior, $(\bar{h}^\circ - \bar{h})$ is zero or close to it.

Liquid phase activity coefficients are complicated to determine because there are deviations from ideal solution behavior. The starting point for evaluating liquid phase activity coefficients is excess free energy,

$$\bar{F}_{\text{excess}} = \bar{F}_{\text{nonideal}} - \bar{F}_{\text{ideal}} \qquad (2.84)$$

This idea together with standard thermodynamics gives the means for determining the activity coefficient. For example, for a system of m components

$$VdP - SdT = n_1 d\mu_1 + n_2 d\mu_2 + \cdots n_m d\mu_m \qquad (2.85)$$

where n represents moles of components and μ is chemical potential for the components.

Next, converting eq 85 to fugacity with mole fractions and applying the result to a binary mixture

$$VdP = x_1 d\mu_1 + x_2 d\mu_2 \qquad (2.86)$$

and

$$(V/RT)dP = x_1 d \ln \bar{f}_1 + x_2 d \ln \bar{f}_2 \qquad (2.87)$$

Using the definition of activity ($\bar{a}_i = f_i/f^0$) and the activity coefficient ($\gamma_i = \bar{a}_i/x_i$), we obtain from eq 87.

$$(V/RT)dP = (d \ln \gamma_1/d \ln x_1) - (d \ln \gamma_2/d \ln x_2) +$$

$$(d \ln f_1^0/d \ln x_1) - (d \ln f_2^0/d \ln x_2) \qquad (2.88)$$

Because

$$d \ln f_i^0/dP = V_i^0/RT$$

then,

$$(V - x_1\ V_1^0 - x_2\ V_2^0)(1/RT)(\partial P/\partial x_1)_T = x_1(d \ln \gamma_1/dx_1) + x_2(d \ln \gamma_2/dx_2) \tag{2.89}$$

The $(V - x_1\ V_1^0 - x_2\ V_2^0)$ term represents the change in volume, that is, V equals $(x_1\ \overline{V}_1 + x_2\ \overline{V}_2)$.

In many situations the left side of eq 89 is negligible. Hence,

$$0 = x_1(d \ln \gamma_1/dx_1) + x_2(d \ln \gamma_2/dx_2) \tag{2.90}$$

and, ultimately,

$$\int_{x_1=0}^{x_1=1.0} \ln \gamma_1/\gamma_2 \ dx_1 = 0 \tag{2.91}$$

This is the Gibbs-Duhem equation for isobaric, isothermal conditions. The equation offers a check of vapor–liquid experimental data consistency. A plot of $\ln \gamma_1/\gamma_2$ versus x_1 for the range zero to one should yield equal areas on either side of the value $\ln \gamma_1/\gamma_2$ zero (Figures 2.16 and 2.17).

In dealing with activity coefficients, we use excess free energy, which is defined as

$$Q = \overline{F}_{excess}/RT = x_1 \ln \gamma_1 + x_2 \ln \gamma_2 \tag{2.92}$$

The Q term is related to a semiempirical relationship involving "interactions" or compositions. The result is that the activity coefficients for the two components of the binary component solution system are now related to the semiempirical constants that evolve from this process. Table 2.5 shows some two-constant equations for binary systems where $\overline{v}_1$ and $\overline{v}_2$ are molal volumes of the pure components and $\overline{v}_1/\overline{v}_{mix}$ and $\overline{v}_2/\overline{v}_{mix}$ are the volume fractions based on pure component volumes. The various constants A_{12}, A_{21}, A'_{12}, and A'_{21} are given by

$$\ln \gamma_1^{\infty} = A_{12} \tag{2.93}$$

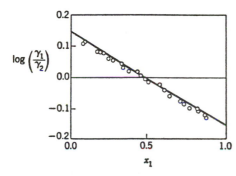

FIGURE 2.16 If the areas above and below the plot of the activity coefficient ratio vs. mole fraction are the same, the experimental data are consistent. The plot represents equation 91 for an n-hexane–toluene system. (Reproduced with permission from reference 12. Copyright 1952 AIChE.)

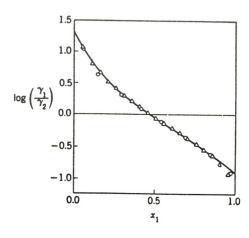

FIGURE 2.17 This plot of the activity coefficient ratio and mole fraction based on experimental data has equal areas above and below the curve, which means that the data are consistent for the system methanol–ethylcyclohexane. The plot represents equation 91. (Reproduced with permission from reference 12. Copyright 1952 AIChE.)

$$\ln \gamma_2^\infty = A_{21} \tag{2.94}$$

$$\ln \gamma_1^\infty = 1 - \ln A'_{12} - A'_{21} \tag{2.95}$$

and

$$\ln \gamma_2^\infty = 1 - \ln A'_{21} - A'_{12} \tag{2.96}$$

where

$$\gamma_i^\infty = \lim_{x_i \to 0} \gamma_i$$

TABLE 2.5 Two-Constant Activity Coefficient Equations

Equation	ln γ_1	ln γ_2
Margules (*13*)	$x_2^2[A_{12} + 2(A_{21} - A_{12})x_1]$	$x_1^2[A_{21} + 2(A_{12} - A_{21})x_2]$
Van Laar (*14*)	$A_{12}[A_{21}x_2/(A_{12}x_1 + A_{21}x_2)]^2$	$A_{21}[A_{12}x_1/(A_{12}x_1 + A_{21}x_2)]^2$
Scatchard–Hamer (*15*)	$[A_{12} + 2(A_{21}\,\bar{v}_1/\bar{v}_2 - A_{12})(\bar{v}_1/\bar{v}_{mix})](\bar{v}_2/\bar{v}_{mix})^2$	$[A_{21} + 2(A_{12}\bar{v}_2/\bar{v}_1 - A_{21})(\bar{v}_2/\bar{v}_{mix})](\bar{v}_1/\bar{v}_{mix})^2$
Wilson (*16*)	$1 - \ln(x_1 + A'_{12}x_2) - \dfrac{x_1}{x_1 + x_2A'_{12}} - \dfrac{A'_{21}x_2}{x_2 + A'_{21}x_1}$	$1 - \ln(x_2 + A'_{21}x_1) - \dfrac{x_2}{x_2 + A'_{21}x_1} - \dfrac{A'_{12}x_1}{x_1 + A'_{12}x_2}$

The two constant equations of Table 2.5 are useful, but we also need relationships with more constants. Among the best known forms are the Redlich–Kister equations (*17*) for activity coefficients

$$\ln \gamma_1 = x_2{}^2 [B + C(3x_1 - x_2) + D(x_1 - x_2)(5x_1 - x_2)] \qquad (2.97)$$

and

$$\ln \gamma_2 = x_1{}^2 [B + C(x_1 - 3x_2) + D(x_1 - x_2)(x_1 - 5x_2)] \qquad (2.98)$$

The various constants B, C, and D are obtained by fitting the $\ln \gamma_1/\gamma_2$ curve versus x_1 or x_2.

Modifications of the preceding activity coefficient equation exist, involving combinations of various binary constants and mole fractions to represent all possible interactions in the multicomponent system. One of the less complicated versions is the multicomponent form of the Wilson equation,

$$\ln \gamma_i = 1 - \ln \left(\sum_j A'_{ij} x_j \right) - \sum_j [(A'_{ji} x_j)/(\sum_k A'_{jk} x_k)] \qquad (2.99)$$

The A_{ij}s are the binary constants given by equations 93 and 94.

The following example illustrates how the Redlich–Kister equation relates activity coefficients with mole fractions. This relationship is of great importance with systems that form nonideal solutions.

EXAMPLE 2.9.   APPLYING THE REDLICH–KISTER EQUATION

At 50 °C, the vapor pressures for ether and ethyl alcohol are 1.7×10^5 and 3.95×10^4 N/m^2, respectively. For the data given below and assuming that the Redlich–Kister relation applies, find the activity coefficients as a function of total pressure.

Mole Fraction Ether	Total Pressure (N/m^2)
0.065	5.35×10^4
0.211	9.29×10^4
0.383	1.21×10^5
0.587	1.44×10^5
0.854	1.62×10^5

Because the total pressures are low,

$$P = \gamma_1 x_1 P_1{}^+ + \gamma_2 x_2 P_2{}^+$$

According to the Redlich–Kister relations (assuming that D is negligible),

$$\ln \gamma_1 = x_2{}^2 [B + C(3x_1 - x_2)]$$

and

$$\ln \gamma_2 = x_1{}^2[B + C(x_1 - 3x_2)]$$

There are three equations with four unknowns (γ_1, γ_2, B, and C). When using the Redlich–Kister equation for low-pressure conditions, we can use a truncated version with only constants B and C, assuming that $B >>> C$. Then

$$\ln \gamma_1 = Bx_2{}^2$$
$$\ln \gamma_2 = Bx_1{}^2$$

and

$$\gamma_1 = e^{Bx_1^2}$$
$$\gamma_2 = e^{Bx_2^2}$$

so that

$$P = (e^{Bx_2^2})\, x_1 P^+ + (e^{Bx_1^2}) x_2 P^+$$

We can then use the ($x_1 - P$) data given to solve this equation for a B value.

$$B = 1.1$$

By trial and error, we can establish the C value that will give a proper fit. The value obtained is C equal to 0.19.

Thus

$$\ln \gamma_1 = x_2{}^2[1.1 + 0.19(3x_1 - x_2)]$$
$$\ln \gamma_2 = x_1{}^2[1.1 + 0.19(x_1 - 3x_2)]$$

The calculated results are then

x_1	γ_1	x_2	γ_2
0.065	2.316	0.935	1.002
0.211	1.947	0.789	1.031
0.383	1.581	0.617	1.108
0.587	1.261	0.413	1.401
0.854	1.037	0.146	2.363

We will discuss phase equilibrium determination again in the mass transfer sections of Chapter 5, which cover separation processes such as distillation, absorption, and extraction.

REFERENCES

1. Nelson, L. C.; Obert, E. F. *Chem. Eng.* (*N.Y.*) **1954,** *16*(7), 203.
2. Keenan, J. H.; Keyes, J. *Thermodynamic Properties of Steam*; John Wiley and Sons: New York, 1936.
3. DuPont, Wilmington, DE.
4. Lydersen, A. L.; Greenkorn, R. A.; Hougen, O. A. *Generalized Thermodynamic Properties of Fluids*; University of Wisconsin: Madison, WI, 1955.
5. Holman, J. P. *Thermodynamics*, 4th ed.; McGraw-Hill: New York, 1988.
6. Benedict, M.; Webb, G. B.; Rubin, L. C. *Chem. Eng. Prog.* **1951,** *47*, 419, 449, 571, 609.
7. M. W. Kellogg Company. *Liquid Vapor Equilibrium in Mixtures of Light Hydrocarbons*; Polyco Data, 1950.
8. De Priester, C. L. *Chem. Eng. Prog. Symp. Ser.* **1953,** 49(7), 53.
9. Edminster, W. C.; Ruby, C. L. *Chem. Eng. Prog.* **1955,** *51*, 95-F.
10. Hadden, S. T. *Chem. Eng. Prog., Symp. Ser.* **1953,** *49*(7), 53.
11. Winn, F. W. *Chem. Eng. Prog. Symp. Ser.* **1952** *48*(2), 121.
12. Redlich, O.; Kister, A. T.; Turnquist, T. E. *Chem. Eng. Prog. Symp. Ser.* **1952,** *48*(2), 49.
13. Margules, M. *Sitzungsber. Akad. Wiss. Wien, Math. Naturwiss. Kl.* **1895,** *104*, 1243.
14. Van Laar, J. J. Z. *Phys. Chem.* **1910,** *72*, 723.
15. Scratchard, G.; Hamer, W. J. *J. Am. Chem. Soc.* **1935,** *57*, 1805.
16. Wilson, G. M. *J. Am. Chem. Soc.* **1964,** *86*, 127.
17. Redlich, O.; Kister, A. T. *Ind. Eng. Chem.* **1958,** *30*, 341, 345.
18. Hougen, O. A.; Watson, K. M. *Chemical Process Principles: Part 2*; John Wiley and Sons: New York, 1947.
19. Smith, J. M.; Van Ness, H. C. *Introduction to Chemical Engineering Thermodynamics*, 4th ed.; McGraw-Hill: New York, 1987.

FURTHER READING

Callen, H. B. *Thermodynamics*, 2nd ed.; John Wiley and Sons: New York, 1985.
Hatsopoulos, G. N.: Keenan, J. N. *Principles of General Thermodynamics*, reproduction ed.; John Wiley and Sons: New York, 1981.
Holman, J. P. *Thermodynamics*, 4th ed.; McGraw-Hill: New York, 1988.
Hougen, O. A.; Watson, K. M.; Ragatz, R. A. *Chemical Process Principles: Part Two, Thermodynamics*, 2nd ed.; John Wiley and Sons: New York, 1959, may be out of print.
Obert, E. F. *Concepts of Thermodynamics*; McGraw-Hill: New York, 1960, out of print.
Reynolds, W. C.; Perkins, H. C. *Thermodynamics*, 2nd ed.; McGraw-Hill: New York, 1977.
Van Wylen, G. J.; Sonntag, R. E. *Fundamentals of Classical Thermodynamics*, 3rd ed.; John Wiley and Sons: New York, 1986.

3

Fluid Flow

Understanding the behavior of fluids is an integral part of engineering in the chemical industry. As a starting point for studying fluid flow, we will first examine the character of fluids at rest (fluid statics) before exploring the nature of fluids in motion (fluid mechanics).

FLUID STATICS

The basic equation of fluid statics, derived from Newton's law of motion, is the barometric equation

$$dP/dZ = \rho g \tag{3.1}$$

where P is the pressure in force per unit area (lbf/ft^2, dyne/cm^2, etc.), Z is the vertical distance or height of the fluid above a reference plane in feet or centimeters, ρ is the fluid density in lb-mass/ft^3 or g/cm^3, and g is the acceleration of the mass due to gravity in ft/s^2 or cm/s^2.

Liquids can exert forces at angles other than 90° to a horizontal surface, so we need to define what the force would be in another direction. To apply the barometric equation to some direction l other than the vertical, we must modify eq 3.1 to

$$dP/dl = \rho g \cos \theta \tag{3.2}$$

where θ is the angle between direction l and the vertical Z axis.

When a fluid has a surface exposed to atmospheric pressure (called a free surface), we can integrate equation 3.1 to obtain the total force of the fluid ($P_2 - P_1$) as

$$P_2 - P_1 = \rho g(Z_2 - Z_1) \tag{3.3}$$

$$(Z_2 - Z_1) = (Z_{\text{free surface}} - Z_1) = h \tag{3.4}$$

Hence

$$(P_2 - P_1) = \rho g h \tag{3.5}$$

but P_1 is atmospheric, so $(P_2 - P_1)$ represents the pressure that a pressure gauge would read, and

$$P_{\text{gauge}} = \rho g h \tag{3.6}$$

where h is the difference in liquid levels of the two sides of a manometer.

Examples 3.1 and 3.2 illustrate pressure measurements encountered in industrial practice that rely on the barometric equation. The approach used in Example 3.1 illustrates the technique that would be used for any system, no matter what its complexity, to determine pressures at any given point or the overall pressure. The reader should also study the discussion of g_c (32.2 lb-mass-ft/lbf-s^2), the factor used to convert mass to force when using English units, following Example 3.1. Understanding how to apply g_c is important because many engineers use the English system of units in calculations.

EXAMPLE 3.1. DETERMINING PRESSURE WITH A MANOMETER

A manometer (Figure 3.1) measures pressure changes $(P_a - P_b)$ in a flowing solution of fluid B with specific gravity 1.26. The manometer fluid (fluid A) is mercury (with a specific gravity of 13.6). If the manometer reading is 332 mm, what is $(P_a - P_b)$?

The pressure measured at various points is

Point 1: $P_1 = P_a$

Point 2: $P_2 = P_a + g(Z_m + R_m)\rho_b$

Point 3: $P_4 = P_a + g(Z_m + R_m)\rho_b - gR_m\rho_a$

Point 5: $P_5 = P_b = P_a + g(Z_m + R_m)\rho_b - gZ_m\rho_b - gR_m\rho_a$

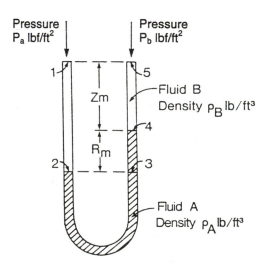

Pressure P_a lbf/ft^2 Pressure P_b lbf/ft^2

Fluid B Density ρ_B lb/ft^3

Fluid A Density ρ_A lb/ft^3

FIGURE 3.1. A manometer is a device used for measuring pressure in industrial processes. For working a sample calculation, see Example 3.1. (Reproduced with permission from reference 8. Copyright 1985 McGraw-Hill.)

Hence,

$$P_b = P_a + g[(Z_m + R_m - Z_m)\rho_b - R_m\rho_a]$$
$$P_b = P_a + g[R_m\rho_b - R_m\rho_a]$$

and

$$P_a - P_b = gR_m(\rho_a - \rho_b)$$

Using the differential height R_m as h

$$P_a - P_b = gh(\rho_a - \rho_b)$$

Substituting numerical values in the equation, we get

$$P_a - P_b = (332 \text{ mm}/304.8 \text{ mm/ft})(32 \text{ ft/s}^2)(13.6 - 1.26)$$
$$(62.4 \text{ lb-mass/ft}^3)/(32.2 \text{ lb-mass-ft/lbf-s}^2)$$
$$P_a - P_b = 833.5 \text{ lbf/ft}^2 = 5.78 \text{ lbf/in}^2 = 39,878 \text{ N/m}^2$$

One of the terms in the above calculation, namely 32.2 lb-mass-ft/lbf-s², is not shown in the relation for $(P_a - P_b)$. This term is known as g_c (see box on p 42).

The g_c represents a conversion of mass and acceleration to force. Units of g_c are (mass)(acceleration)/(force). The conversion g_c must be used for all unit systems, including SI and cgs, where g_c is equal to unity. If it is not included, units of force will not appear in calculations.

In the English system of units, g_c is 32.2 lb-mass-ft/lbf · s² (Table 3.1).

EXAMPLE 3.2. STATIC PRESSURE IN A TANK

A system (Figure 3.2) has a manometer filled with ethyl iodide (specific gravity of 1.93). The heights h_1 and h_2 are 1.12 and 0.20 m, respectively. What are the gauge and absolute pressures in the tank?

$$P_{gauge} = (P_a - P_b) = gh(\rho_a - \rho_b)$$
$$P_{gauge} = (1.93)(998 \text{ kg/m}^3)(1.32 \text{ m})(9.81 \text{ m/s}^2) = 2.50 \times 10^4 \text{ N/m}^2$$

TABLE 3.1. Values of g_c for Various Unit Systems

Mass	Length	Time	g_c	Force
Gram	centimeter	second	1 gm-cm/s²-dyne	dyne
Kilogram	meter	second	1 kg-m/s²-newton	newton
Pound mass	foot	second	32.2 lb-mass-ft/lbf-s²	pound force
Slug	foot	second	1 slug-ft/lbf-s²	pound force

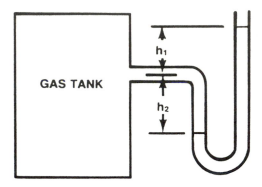

FIGURE 3.2. Gauge pressure is measured by a manometer with one leg open to atmospheric pressure. The total difference in height of the ethyl iodide which fills this manometer (see Example 3.2) is the sum of h_1 and h_2. (Reproduced with permission from reference 9. Copyright 1970 McGraw-Hill.)

To find the absolute pressure, add atmospheric pressure.

$$P_{\text{absolute}} = (2.50 + 10.14)(10^4) \text{ N/m}^2 = 1.264 \times 10^5 \text{ N/m}^2$$

BUOYANCY

Buoyancy is a characteristic of a static fluid frequently calculated in industrial processes. The buoyant force for an object floating in a liquid is

$$F_{\text{buoyant}} = (\rho_{\text{liquid}} g V_{\text{liquid}} + \rho_{\text{air}} g V_{\text{air}}) \tag{3.7}$$

where V_{liquid} is the volume of liquid displaced by the object, and V_{air} equals the volume of air displaced.

EXAMPLE 3.3. CALCULATING BUOYANCY

A participant in a cross-country balloon race has developed a balloon with a highly flexible skin of negligible weight and is considering whether to use hydrogen or helium to maximize the balloon's lifting capacity. Which gas will offer the greatest lifting force and by how much?

$$\text{lifting capacity} = \text{buoyant force} - \text{balloon weight}$$

$$\text{lifting capacity} = V_g(\rho_{\text{air}} - \rho_{\text{gas}})$$

From this equation, we can see that hydrogen, which is less dense than helium, has greater lifting capacity.

The increase in lifting capacity as a percent increase is

$$\% \text{ increase} = \left(\frac{V_g(\rho_{\text{air}} - \rho_{\text{hydrogen}}) - V_g(\rho_{\text{air}} - \rho_{\text{helium}})}{V_g(\rho_{\text{air}} - \rho_{\text{helium}})} \right) 100$$

$$\% \text{ increase} = \left(\frac{(\rho_{\text{air}} - \rho_{\text{hydrogen}})}{(\rho_{\text{air}} - \rho_{\text{helium}})} - 1 \right) 100$$

Assuming 70 °F and 1 atmosphere and using the ideal gas law to obtain densities,

$$\% \text{ increase} = ((0.075 - 0.005)/(0.075 - 0.010) - 1) \, 100 = 7.7\%$$

FLUID DYNAMICS—THE CHARACTER OF FLOWING FLUIDS

When a fluid is in motion, its behavior is determined by its physical nature, the flow geometry (the shape of the channel in which the fluid flows), and the flow velocity. Many years ago, Osborne Reynolds conducted a series of experiments exploring the nature of flow. One of the experiments consisted of injecting a dye stream into water flowing in a tube (Figure 3.3).

WHAT REYNOLDS FOUND

At low flow velocities, Reynolds found that the dye stream forms a smooth, straight streak. As flow velocity increases, the dye becomes sinuous, and at higher velocities it completely disperses. At low velocity the fluid flow is streamline and without mixing perpendicular to the tube axis. Because the fluid appears to move in shells or lamina, the flow is called laminar flow. At high flow rates, there is considerable radial mixing, and the flow pattern is chaotic. This flow is called turbulent. Flows that oscillate between laminar and turbulent flow are called transition flows.

Reynolds further delineated the differences between laminar and turbulent flow by studying the flow velocity profiles in a circular tube for laminar and turbulent conditions. In laminar flow, the velocity profile is parabolic, with zero velocity at the tube wall and maximum velocity in the center (Figure 3.4 for velocity profiles). In contrast, the profile for turbulent flow is blunter. In highly turbulent flow, all velocities across the tube are nearly the same. The ultimate flow is called plug or ideal flow.

The fluid velocity profiles and characteristics described by Reynolds influence the design of tubular reactors. A highly turbulent flow pattern is desired for tubular flow reactors because all portions of the fluid have similar velocities with similar residence times and because the vortices and eddies in turbulent flow effectively mix the fluid. The least desirable pattern is laminar flow because of its stratification of the fluid layers and wide variation in residence times, neither of which contribute to uniform chemical reaction.

THE REYNOLDS NUMBER—A DIMENSIONLESS COMPARISON OF FLOWS

Types of tubular flow behavior can be described by observing and comparing their principal forces—inertial ($\rho \overline{V}^2$) and viscous ($\mu \overline{V}/D$)—where ρ is fluid density, $\overline{V}$ is the average velocity, μ is the viscosity, and D is the tube diameter. Comparing the

FIGURE 3.3. A dye released into laminar and turbulent flow in a tube, an experiment conducted by Osborne Reynolds and called the Reynolds experiment, reveals different velocity profiles. (Reproduced with permission from reference 11. Copyright 1964 Butterworth–Heinemann.)

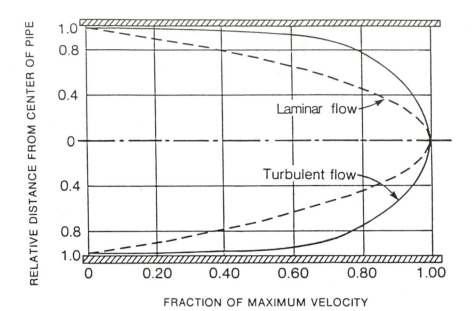

FIGURE 3.4. Laminar and turbulent flows have different velocity distributions as measured across the pipe. (Reproduced with permission from reference 8. Copyright 1985 McGraw-Hill.)

inertial forces to the viscous forces, we obtain a dimensionless group called the Reynolds number (Re).

$$\text{Re} = \frac{(\rho \overline{V}^2)}{} = \frac{D\overline{V}\rho}{\mu} \tag{3.8}$$

Below a Reynolds number of 2100, viscous forces predominate and flow is laminar. From Reynolds numbers of 2100 to 4000, flow is intermediate or transitional. A Reynolds number greater than 4000 indicates that flow is turbulent.

As mentioned in Chapter 1, dimensionless groups of numbers, such as the Re, are useful when comparing processes at different scales.

FLOW AND THE FIRST LAW OF THERMODYNAMICS— BERNOULLI'S EQUATION

To understand intricate flow patterns, we need detailed flow studies. However, it is possible to simplify the solution of flow problems and avoid the minute flow details if we consider flows on an overall basis. The starting point for an overall view is the first law of thermodynamics applied to a flowing system (Figure 3.5) and covered in Chapter 2:

$$-dW_s/dm - (\Delta u - dQ/dm) = \Delta(P/\rho + gZ + \overline{V}^2/2) \tag{3.9}$$

where P is pressure, ρ is density, Z is height above a datum plane, $\overline{V}$ is the average velocity of flow, W_s is shaft work (all work other than injection work, i.e., mechan-

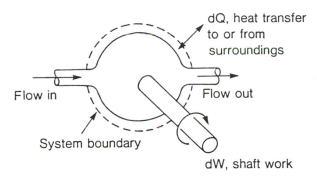

FIGURE 3.5. The first law of thermodynamics can be applied to flow systems by drawing a boundary around the area of interest (shown by the dotted line for this pump) and balancing the energy across the system. The energy includes shaft work, heat transfer to or from the surrounding area, and energy in the flow. (Reproduced with permission from reference 9. Copyright 1970 McGraw-Hill.)

ical work), u is the internal energy change, Q is heat, and m is mass. The equation basically balances injection work $(\Delta P/\rho)$, potential energy $(g\Delta Z)$, and kinetic energy $(\Delta \overline{V}^2/2)$ against shaft work, internal energy, and heat.

Equation 3.9 for an incompressible fluid (a good approximation for most liquids and also for gases under certain conditions) equates the change of internal energy and heat to a friction heating term

$$(\Delta u - dQ/dm) = F_h \tag{3.10}$$

Friction heating can best be understood by realizing that the energy required to move a fluid is converted to a non-useful form.

Modifying the conservation of energy equation for fluid friction enables us to rewrite eq 3.9 as

$$\frac{-dW_s}{dm} - F_h = \Delta\left(\frac{P}{\rho} + gZ + \frac{\overline{V}^2}{2}\right) \tag{3.11}$$

Equation 3.11 is called the Bernoulli equation or the mechanical energy balance.

When the Bernoulli equation is divided by g, all terms in the resulting equation

$$\frac{-dW_s}{dm}\left(\frac{1}{g}\right) - \frac{F_h}{g} = \Delta\left(\frac{P}{\rho g} + Z + \frac{\overline{V}^2}{2g}\right) \tag{3.12}$$

are in the form of a length or "head", which is the driving force for fluid flows in open channels and dams (the flows are at atmospheric pressure). Engineers frequently refer to flows in terms of "head", meaning the height of a tank or source of flow.

General application of the Bernoulli equation requires a value for the friction heating term, F_h. However, there are a number of cases where friction losses can be neglected, as in the following examples. Each example problem is solved by writing the complete Bernoulli balance first and then canceling or neglecting terms appropriately. The final form of the equation reveals the important contributions to the physical process.

EXAMPLE 3.4. FLOW DUE TO HEAD PRESSURE

The tank shown in Figure 3.6 has a cross-sectional exit area of 0.279 m^2 and a height h of 3 m. If flow from the tank is frictionless, what is the volumetric flow rate?

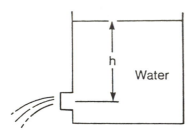

FIGURE 3.6. Fluid flow from the pipeline of a tank (see Example 3.4) represents a conversion of potential to kinetic energy. (Reproduced with permission from reference 9. Copyright 1970 McGraw-Hill.)

For this case, both the inlet and the outlet of the tank are at atmospheric pressure, no mechanical shaft work is performed, and there is no friction, so

$$-\frac{dW_s}{dm} - F_h = \Delta\left(\frac{P}{\rho} + gZ + \frac{\overline{V^2}}{2}\right)$$

becomes

$$0 = \Delta\left(gZ + \frac{\overline{V^2}}{2g}\right)$$

The tank diameter is large with respect to the diameter of the outflow tube. Thus the velocity V_1 of the fluid at the tank top is negligible compared to the velocity V_2 at outflow.

$$0 = g(Z_2 - Z_1) + \left(\frac{\overline{V_2}^2}{2} - \frac{\overline{V_1}^2}{2}\right)$$

$$0 = -gh + \overline{V_2}^2/2$$

$$\overline{V_2} = (2gh)^{1/2}$$

$$\overline{V_2} = [2(9.81 \text{ m/s}^2)(3 \text{ m})]^{1/2}$$

$$\overline{V_2} = 7.67 \text{ m/s}$$

and the volumetric flow rate equals

$$(7.67 \text{ m/s})(0.279 \text{ m}^2) = 2.14 \text{ m}^3/\text{s} = 75.6 \text{ ft}^3/\text{s}$$

EXAMPLE 3.5. FLOW DUE TO DENSITY DIFFERENCES

A tank of water is placed in a tank of gasoline (Figure 3.7). What is the velocity of water at the outflow?

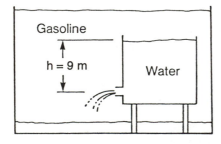

FIGURE 3.7. Calculating the flow of water from a tank immersed in gasoline must take into account both pressure and potential to kinetic energy changes (see Example 3.5). (Reproduced with permission from reference 9. Copyright 1970 McGraw-Hill.)

Because there is no shaft work and no friction, the Bernoulli equation becomes

$$0 = \Delta(P/\rho + gZ + \overline{V}^2/2)$$

The tank diameter is large compared to outflow tube diameter, and thus $\overline{V}_2 >>>> \overline{V}_1$. Then

$$\overline{V}_2^2/2 = -\Delta P/\rho_{\text{water}} - g\Delta Z$$

$$\overline{V}_2 = [2|(-\Delta P/\rho_{\text{water}}) - g\Delta Z|]^{1/2}$$

$$\Delta P = (\rho_{\text{gasoline}})g\Delta Z$$

and

$$\overline{V}_2 = [-2g\Delta Z(\rho_{\text{gasoline}} - \rho_{\text{water}})/\rho_{\text{water}}]^{1/2}$$

The density terms can be written in terms of specific gravities as

$$\frac{0.72 - 1.0}{1.0}$$

Then,

$$\overline{V}_2 = [-(2)(9.81 \text{ m/s}^2)(9 \text{ m})(0.72 - 1)]^{1/2} = 7.03 \text{ m/s} = 23.1 \text{ ft/s}$$

FRICTIONLESS FLOW MEASUREMENT DEVICES

The frictionless form of the Bernoulli equation is the basis for many fluid flow measuring devices, including the pitot tube, the pitot-static tube, the venturi meter, and the orifice meter (Figure 3.8).

The frictionless Bernoulli equation is

$$(P_2 - P_1)/\rho + (\overline{V}_2^2 - \overline{V}_1^2)/2 = 0 \tag{3.13}$$

which ignores friction, height effects, and shaft work.

For the pitot tube (a tube inserted below the surface of a moving fluid (Figure 3.8A), the equation is

$$\overline{V}_1 = (2gh_1)^{1/2} \tag{3.14}$$

Likewise, for the pitot-static tube (Figure 3.8B), the equation is

$$\overline{V}_1 = (2\Delta P/\rho)^{1/2} \tag{3.15}$$

The venturi meter (Figure 3.8C) is handled by

$$\overline{V}_2 = [(2)(P_1 - P_2)/\rho(1 - A_2^2/A_1^2)]^{1/2} \tag{3.16}$$

where A_1 is the area of the meter opening at point 1 and A_2 is the area of the throat at point 2. However, the flow rate calculated from eq 3.16 is higher than that observed due to losses from two factors, frictional heating and nonuniform flow. To

A. Pitot tube

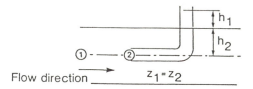

B. Pitot-static tube

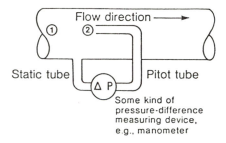

C. Venturi meter

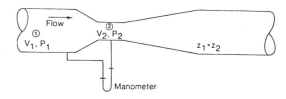

FIGURE 3.8. Bernoulli's equation is the basis of calculations for all of the flow meters shown here. The pitot tube (A) measures the change in energy from kinetic to potential. The pitot-static tube (B) measures the difference between the static pressure and the kinetic energy of a flow. The venturi meter (C), which creates a pressure drop in the flow, is very accurate but expensive. The orifice meter (D) creates a pressure drop in the flow with an orifice plate of specific diameter. (Reproduced with permission from reference 9. Copyright 1970 McGraw-Hill.)

D. Orifice meter

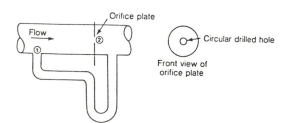

compensate for these losses, an empirical coefficient C_V, the discharge coefficient, is introduced into eq 3.17:

$$\overline{V}_2 = C_V \sqrt{[2(P_1 - P_2)/\rho]/(1 - A_2{}^2/A_1{}^2)} \tag{3.17}$$

Values of C_V versus the Reynolds number are presented in graphical form in Figure 3.9 for the venturi meter.

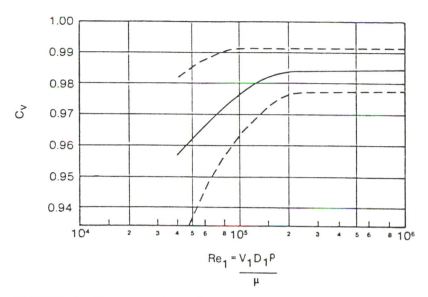

FIGURE 3.9. Velocity calculations for a venturi meter are corrected with a coefficient of discharge C_V based on the Reynolds number. The solid line is the data line. Dotted lines indicate the range of deviation. (Reproduced with permission from reference 12. Copyright 1959 ASME International.)

The venturi meter is a reliable, high-quality, but complex and costly flow meter. A simpler and less expensive device is the orifice meter (Figure 3.8D). Eq 3.17 can be applied to an orifice meter by using C_V values from Figure 3.10.

The following example illustrates the application of the orifice meter and underscores the lack of information that hinders proper analysis in many engineering calculations. However, engineering problem solving overcomes this difficulty with appropriate insights.

EXAMPLE 3.6. MEASURING FLOW WITH AN ORIFICE PLATE

What is the velocity measured by an orifice plate hole 0.06 m in diameter in a 0.30-m diameter pipe, with a measured pressure drop of 7.51×10^4 N/m²?

$$\overline{V}_2 = C_V \sqrt{[2(P_1 - P_2)/\rho]/(1 - A_2{}^2/A_1{}^2)}$$

All quantities except $\overline{V}_2$ and C_V are known. However, we need to know the velocity of the fluid in the pipe $\overline{V}_1$ to find C_V from Figure 3.10. This velocity is unknown and cannot be found until $\overline{V}_2$ is determined. Without C_V we cannot find $\overline{V}_2$, which in turn gives $\overline{V}_1$ needed to find C_V.

This situation is typical in engineering calculations, requiring a trial and error solution. C_V for a diameter ratio of 0.20 varies from 0.60 to 0.70. Starting with a 0.65 value,

$$\overline{V}_2 = 0.65 \sqrt{[2(7.51 \times 10^4 \text{ N/m}^3)]/[(998 \text{ kg/m}^3)(1 - \pi^2 D_2{}^4/\pi^2 D_1{}^4)]}$$

$$\overline{V}_2 = 7.98 \text{ m/s}$$

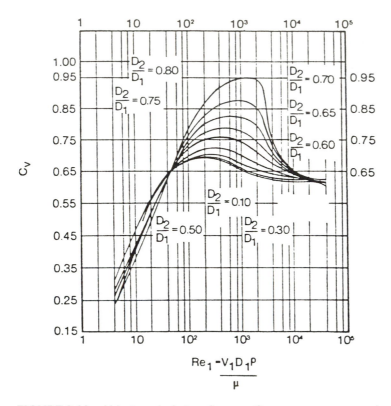

FIGURE 3.10. Velocity calculations for an orifice meter are corrected with a coefficient of discharge C_V correlated to the Reynolds number. D_1 is the diameter of the pipe, and D_2 is the diameter of the orifice plate opening. (Reproduced with permission from reference 13. Copyright 1933 Chilton's Instrumentation and Control Systems.)

and

$$\overline{V}_1 = \overline{V}_2 A_2/A_1 = (7.98 \text{ m/s})(0.06 \text{ m})^2/(0.30 \text{ m})^2 = 0.32 \text{ m/s}$$

The estimated $\overline{V}_1$ will not give a 0.65 C_V value but rather 0.625. Using the new C_V value to re-calculate $\overline{V}_2$ gives a $\overline{V}_1$ of 0.30 m/s and a Re_1 of 18,000. This agrees with the new, assumed C_V value.

FRICTION LOSSES IN FLUID FLOW

There are instances where frictional forces F_h can be neglected in the Bernoulli equation. However, there are many cases where the F_h term must be evaluated in long pipe runs; runs with many elbows, bends, and tees; and runs with sudden expansions and contractions. For an accurate idea of the pressure and flow possible in these processes, we must account for friction.

FRICTION IN LAMINAR FLOW

The F_h term in laminar flow can be determined directly by setting up a force balance on flow in a tube (Figure 3.11) between the pressure and shear forces.

$$\text{shear force} = (2r\pi L)(\text{shear stress at radius } r) \tag{3.18}$$

$$\tau = (\text{shear stress at radius } r) = r(P_1 - P_2)/2L \tag{3.19}$$

Shear stress increases with distance from the center of the pipe (Figure 3.12). Although this shear stress relationship is derived for laminar flow, it also applies to turbulent flow and to the non-Newtonian fluids we will consider later.

From Newton's law

$$\tau = -\mu \, dV/dr \tag{3.20}$$

Equations 3.19 and 3.20 are expressions for τ. Equating them,

$$dV = \int -\frac{r(P_1 - P_2)}{\mu 2L} \, dr \tag{3.21}$$

Integrating from the pipe wall ($r = r_o$) to the center of the pipe at r, and knowing that velocity V equals zero at the pipe wall,

$$V = (r_o^2 - r^2)(P_1 - P_2)/4 \, \mu L \tag{3.22}$$

For volumetric flow (Q)

$$Q = \int V \, dA = \left(\frac{P_1 - P_2}{L}\right)\left(\frac{D^4}{128}\right)\left(\frac{\pi}{\mu}\right) \tag{3.23}$$

where

$$-\Delta P = (P_1 - P_2) = \frac{QL \, \mu \, 128}{\pi \, D^4}$$

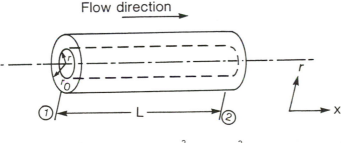

Flow direction

Pressure force $= P_1(\pi r^2) - P_2(\pi r^2)$

FIGURE 3.11. Two forces are considered when examining the frictional heating that occurs due to flow, the pressure force on the cross-sectional area and the shear force on the surface area. (Reproduced with permission from reference 9. Copyright 1970 McGraw-Hill.)

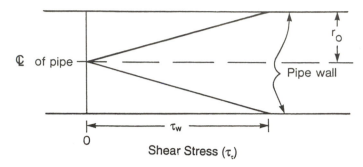

FIGURE 3.12.   Shear stress is a function of radius. (Reproduced with permission from reference 8. Copyright 1985 McGraw-Hill.)

Applying Bernoulli's equation to the tube and assuming no change in height Z or kinetic energy and no shaft work,

$$-\frac{dW_s}{dm} - F_h = \frac{\Delta P}{\rho} + g\Delta Z + \frac{\Delta \bar{V}^2}{2}$$

and

$$F_h = -\Delta P/\rho \tag{3.24}$$

Substituting the value of ΔP from eq 3.23 in eq 3.24,

$$-\Delta P/\rho = QL\mu \, 128/\rho\pi D^4 \tag{3.25}$$

so that

$$F_h = (QL)(\mu/\rho)(128/\pi D^4) \tag{3.26}$$

This expression for friction applies to laminar flow in a tube, whether flow is vertical or on a slope.

FRICTION IN TURBULENT FLOW

The chaotic pattern of turbulent flow makes it difficult to correlate frictional heating with velocity. Earlier, we showed that laminar flow becomes sinuous as velocity increases and then breaks into whirlpools. In turbulent flow the fluid moves along the axis of a tube and in other directions. The net result is a varying velocity $\bar{v}_x$ at a given point, with a time-smoothed velocity $\bar{v}_x$ (Figure 3.13).

The true velocity is the time-smoothed velocity $\bar{v}_x$ plus the fluctuating velocity v_x'. If we substituted this combination of velocities in the complicated differential flow equations, we would have to time-smooth the overall relationships. When we do this, we obtain additional stresses, called Reynolds stresses, that do not appear in the laminar flow. These additional stresses prevent an analysis similar to that for laminar flow. Therefore, a semiempirical approach must be taken to analyze frictional forces in turbulent flow.

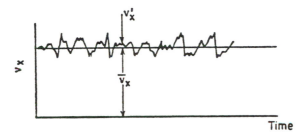

FIGURE 3.13. In turbulent flow, the velocity v_x fluctuates around a time-smoothed velocity. (Reproduced with permission from reference 14. Copyright 1961 Prentice-Hall.)

Friction is proportional to the stress in the fluid, which is a function of the average velocity gradient $(\overline{V}/D)$ and the average velocity $(\overline{V})$ itself. Hence

$$F_h \approx \overline{V}^2/D \qquad (3.27)$$

where $\overline{V}$ is average fluid velocity and D is the diameter of the tube.

F_h is proportional to the tube length, L.

$$F_h \approx L\overline{V}^2/D \qquad (3.28)$$

F_h is not directly equal to $L\overline{V}^2/D$ but proportional to it. Inserting an empirical function f, the friction factor, into eq 3.28,

$$F_h = \frac{2fL\overline{V}^2}{g_c D} \qquad (3.29)$$

Experiments on different tubes and pipes provide values for the function f. For tubes made of smooth material, such as glass or brass, f is a function of the Reynolds number,

$$f = \phi\left(\frac{D\overline{V}\rho}{\mu}\right) = \phi(\text{Re}) \qquad (3.30)$$

Tubes can have rough surfaces that influence friction factors (Figure 3.14); the surfaces represent idealized versions of rough tubes. For rough tubes, the friction factor is a function of the Reynolds number and the ratio of the depth of the surface roughness ϵ to the tube diameter,

$$f = \phi\left(\frac{D\overline{V}\rho}{\mu}, \frac{\epsilon}{D}\right) = \phi\left(\text{Re}, \frac{\epsilon}{D}\right) \qquad (3.31)$$

The protuberances can have shapes different from those shown (Figure 3.14); ϵ is an average value. Table 3.2 summarizes the value of ϵ for pipes used in industrial processes.

The overall correlation of f with the Reynolds number and the roughness factor ϵ/D is plotted graphically in Figure 3.15.

Behavior in the laminar region is correlated with friction by the expression

$$f = 16/\text{Re} \qquad (3.32)$$

which means that tube roughness does not affect F_h in laminar flow.

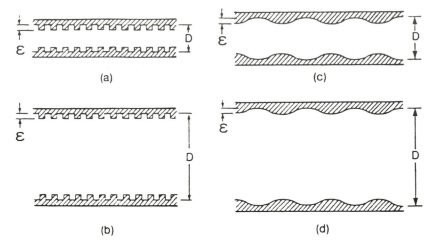

FIGURE 3.14. Most tubes have some roughness, which effects a pressure drop with flow due to friction. The roughness varies with the configuration of the pipe surface. ϵ is the average height of the protuberances. (Reproduced with permission from reference 8. Copyright 1985 McGraw-Hill.)

There are two correlations between f and Re that are used by engineers. The one in Figure 3.15 is used by chemical engineers and applies the laminar flow relation of eq 3.32, namely, $f = 16/\text{Re}$. The other correlation is used by civil engineers and has a laminar relation of $f = 64/\text{Re}$. The two values for f are related by a factor of 4. The reader should check the laminar flow region of any f–Re chart to determine which one is being used.

Straight tubing and piping surfaces are only one source of frictional loss. Other sources of friction and pressure loss include elbows, tees, valves, and pipe unions (Figure 3.16). For these fittings, we assume that frictional heating due to flow is a constant expressed as an equivalent length of straight pipe for a given diameter

$$F_{h \text{ fitting}} = (\text{constant for fitting})(F_h \text{ for a pipe length equal to one pipe diameter})$$

$$(3.33)$$

TABLE 3.2. Surface Roughness for Various Materials

Material	Surface Roughness (ϵ cm)
Drawn tubing (brass, lead, glass, etc.)	0.00015
Commercial steel or wrought iron	0.0046
Asphalted cast iron	0.0122
Galvanized iron	0.0152
Cast iron	0.0259
Wood stave	0.0183–0.0914
Concrete	0.0305–0.3048
Riveted steel	0.0914–0.9144

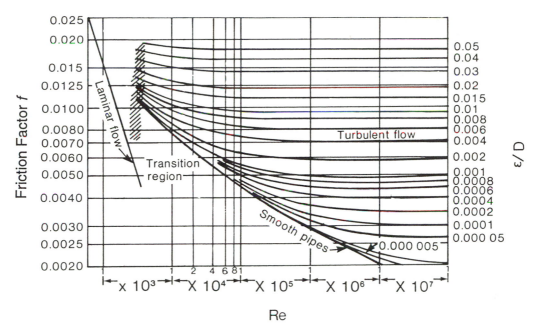

FIGURE 3.15. The friction factor *f* varies with the Reynolds number and possibly roughness, as shown by this friction factor correlation for pipes.

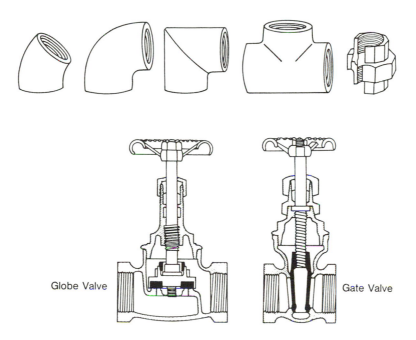

Globe Valve   Gate Valve

FIGURE 3.16. Pipe fittings and valves, such as these elbows, tees, globe valve, and gate valve, create pressure drops in piping and must be considered as additional lengths of piping in flow calculations. (Reproduced with permission from reference 11. Copyright 1964 Butterworth–Heinemann.)

TABLE 3.3.   Equivalent Lengths of Piping for Fittings

Type of Fitting	Equivalent Length L/D (dimensionless)
Globe valve, wide open	340
Angle valve, wide open	145
Gate valve, wide open	13
Check valve, swing type	135
90° Standard elbow	30
45° Standard elbow	16
90° Long-radius elbow	20

In pressure drop calculations, we add a frictional equivalent length of piping for each fitting in the system. Equivalent lengths are listed in Table 3.3 for various fittings.

Sudden expansions and contractions also cause frictional heating (Figures 3.17 and 3.18). Enlargements can have eddylike flow in the corners of the larger duct (*see* Figure 3.17, left figure). Contractions (Figure 3.17, right figure, and Figure 3.18) create a boundary layer of fluid at the wall. A velocity profile develops within the layer and joins with the main flow to form a complete cross-tube profile. The axial length for this profile development is called the *entrance length.*

Because frictional heating is complex, we use the concept of kinetic energy to describe it

$$F_h = \frac{K\overline{V}^2}{2} \tag{3.34}$$

where K values are shown as functions of diameter ratios in Figure 3.19.

Friction is also caused by flow through a conduit that is not circular, for example, flow between parallel plates, in annuli, etc. For laminar flow between parallel plates, the relationship between the volumetric flow rate Q and pressure P is

$$Q = \frac{(P_1 - P_2)}{L} \frac{1}{\mu} \frac{1}{12} wh \tag{3.35}$$

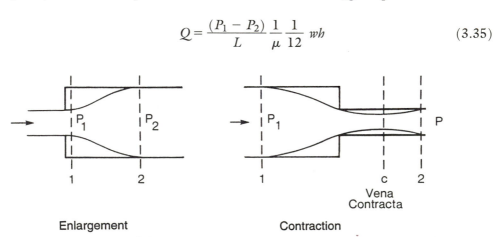

Enlargement   Contraction

FIGURE 3.17.   Sudden enlargement or contraction of the flow area must be considered as sources of pressure drop in flow calculations. Note that, with a sudden change in area, the pressure does not recover immediately. (Reproduced with permission from reference 11. Copyright 1964 Butterworth–Heinemann.)

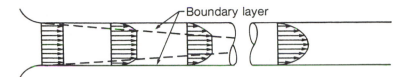

FIGURE 3.18. After a contraction, a boundary layer of fluid forms at the wall. The boundary layer forms its own profile, which gradually merges with the overall flow profile. (Reproduced with permission from reference 8. Copyright 1985 McGraw-Hill.)

where h is the distance between plates and w is the width of the opening between the plates.

Likewise, in laminar flow through an annulus,

$$Q = \frac{(P_1 - P_2)}{L} \frac{1}{\mu} \frac{\pi}{128} (D_0{}^2 - D_1{}^2) \left[(D_0{}^2 - D_1{}^2) - \left(\frac{(D_0{}^2 - D_1{}^2)}{\ln \frac{D_0}{D_1}} \right) \right] \quad (3.36)$$

where D_0 is the outer and D_1 the inner diameter of the annulus.

For turbulent flow through a noncircular conduit, we approach the solution as we did for flow through tubes, by examining the shear stresses. We assume that the

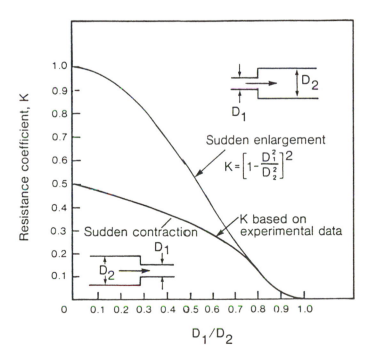

FIGURE 3.19. Sudden expansions and contractions in the flow area create a resistance K to flow that correlates with the ratio of the diameters D_1/D_2 of the areas. (Reproduced with permission from reference 9. Copyright 1970 Crane Valves.)

wall shear stress will be the same for a given average velocity regardless of conduit shape. We define a new term, the hydraulic radius R_h as

$$R_h = \frac{\text{cross-sectional area perpendicular to flow}}{\text{wetted perimeter}} \qquad (3.37)$$

Then,

$$\left(\frac{\Delta P}{L}\right)_{\text{noncircular}} = \frac{f\rho \bar{V}^2}{2R_h} \qquad (3.38)$$

For a pipe, R_h is equal to $D/4$, and eq 3.38 becomes the expression for the circular conduit.

$$\left(\frac{\Delta P}{L}\right)_{\text{circular}} = \frac{2f\rho \bar{V}^2}{D} \qquad (3.39)$$

We can, therefore, replace D by $4R_h$ in turbulent flow. Equation 3.38 can also be used for laminar flow with complex cross sections, if we recognize that the results will be on the high side.

The next two examples illustrate the calculation of frictional heating as a way to compare pressure drops per unit length. They also show what happens to frictional heating in the transition zone between laminar and turbulent flow.

EXAMPLE 3.7. CALCULATING LAMINAR AIR FLOW

Air is flowing through a horizontal tube of 2.54 cm diameter. What is the maximum average velocity at which laminar flow will be stable? What is the pressure drop at this velocity?

The transition point to unstable flow occurs at a Reynolds number of 2100. Hence

$$\frac{D\bar{V}\rho}{\mu} = 2100$$

The values for ρ and μ for air can be obtained from a handbook. However, for less common gases or liquids, the reader can use the appendixes in this book or a source such as reference 1.

$$\bar{V} = \frac{2100\,\mu}{D\rho} = (2100)(1.8 \times 10^{-5}\text{ kg/m-s})/(0.0254\text{ m})(1.2\text{ kg/m}^3)$$
$$\bar{V} = 1.24\text{ m/s}$$
$$\Delta P/L = 128\ Q\mu/D^4\pi = 32\ \bar{V}\ \mu/D^2$$
$$\Delta P/L = (32)(1.24\text{ m/s})(1.8 \times 10^{-5}\text{ kg/m-s})/(0.0254\text{ m})^2$$

and

$$\Delta P/L = 1.107\text{ N/m}^2/\text{m}$$

EXAMPLE 3.8. CALCULATING THE PRESSURE DROP OF LIQUID FLOW

Two large open reservoirs are connected by 1524 m (5000 ft) of 0.20-m-diameter pipe (Figure 3.20). The level in one reservoir is 60.96 m (200 ft) above the level in the other. What is the pressure drop per foot of pipe?

Starting with the Bernoulli equation

$$\frac{-dW_s}{dm} - F_h = \Delta\left(\frac{P}{\rho} + gZ + \frac{V^2}{2}\right)$$

and assuming no change in pressure due to density, no change in kinetic energy, and no shaft work, we get

$$-F_h = g\Delta Z$$

$$-F_h = g(Z_2 - Z_1) = (9.81 \text{ m/s}^2)(60.96 \text{ m})$$

$$F_h = 597.41 \text{ m}^2/\text{s}^2$$

and

$$-\Delta P/L = \rho F_h/L = (999.6 \text{ kg/m}^3)(597.41 \text{ m}^2/\text{s}^2)/(1524 \text{ m})$$

$$-\Delta P/L = 391.84 \text{ N/m}^2/\text{m}$$

Sizing a pump, the topic of the next example, is a typical engineering problem in industry and a classic application of the Bernoulli equation. Even though the methods of estimating frictional heating losses are semiempirical, we could not size the pump without them. The friction factor correlation provides a compact database for any and all types of conduit flow for Newtonian fluids (i.e., for the full range of Reynolds numbers, for all fluid densities, for both gases and liquids, for all pipe diameters, and for all viscosities).

EXAMPLE 3.9. SIZING A PUMP

A pump moves water at 10 °C from a large open reservoir to the bottom of an open elevated tank (Figure 3.21). The level of the tank averages 48.77 m above the surface of the reservoir. The pipe is 0.076 m in diameter, and the line to the tank is 152.4 m of straight pipe, six elbows, two gate valves, and 2 tees ($L/D = 60$). The pump delivers 0.00898 m^3/s. What is the power consumed if the pump has a mechanical efficiency of 55%?

From the Bernoulli equation, taking the two liquid surfaces as boundaries and

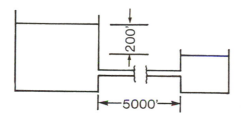

FIGURE 3.20. A difference in liquid levels (the heads) between two tanks creates a flow that will lose pressure from friction (see Example 3.8).

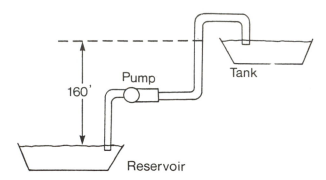

Pump

Tank

160'

Reservoir

FIGURE 3.21. Calculating the size of a pump needed to move liquid from one area to another is a standard engineering task (see Example 3.9).

noting that there is no essential change in atmospheric pressure between the tanks and no change in the kinetic energy,

$$-dW_s/dm - F_h = g\Delta Z$$

and

$$F_h = \frac{2f L \bar{V}^2}{g_c D}$$

Next we calculate the amount of equivalent piping, using the L/D values from Table 3.3 for the fittings. L is the fitting equivalent length, so that

$$L = 152.4 \text{ m} + (0.076 \text{ m})[(30)(6) + (2)(13) + (2)(60)] = 177.2 \text{ m}$$

Next we need the velocity, calculated from the pump's volumetric flow rate,

$$\bar{V} = (0.00898 \text{ m}^3/\text{s})(4)/(\pi)(0.076 \text{ m})^2 = 1.98 \text{ m/s}$$

To estimate the friction factor, we compute the Reynolds number,

$$\text{Re} = D\bar{V}\rho/\mu = (0.076 \text{ m})(1.98 \text{ m/s})(999.6 \text{ kg/m}^3)/(1.3 \times 10^{-3} \text{ kg/m-s})$$
$$\text{Re} = 115,700$$

and from Figure 3.15, $f = 0.0046$.

$$-dW_s/dm = [(4)(0.0046)(177.2 \text{ m})(1.98 \text{ m/s})^2/(2)(0.076 \text{ m})$$
$$+ (48.77 \text{ m})(9.8 \text{ m/s}^2)][1/\text{kg-m/N-s}^2]$$
$$-dW_s/dm = 0.562 \text{ kJ/kg}$$

If g_c (i.e., kg-m/N-s^2) was not included in the calculation of $-dW_s/dm$, the resultant units would have been m^2/s^2, not units of energy per unit mass.

$$(-dW_s/dm)(dm/dt) = (0.562 \text{ kJ/kg})(0.00898 \text{ m}^3/\text{s})(999.6 \text{ kg/m}^3)$$
$$\text{power} = 5.05 \text{ kW}$$

and the total power consumed is 5.05 kW/0.55 or 9.17 kW.

TYPES OF PUMPS AND THEIR USES

The preceding example shows how a pump can be sized for a given system, but there are other considerations for final pump selection. The chemical nature of the fluid (corrosive, flammable, etc.) dictates the materials of construction required in the pump. The physical nature of the fluid, particularly the viscosity, is also important. The following descriptions of pump types explain the various pumps that are available.

Centrifugal pumps are simple in construction, economical, have no valves, operate at high speed with a steady delivery, require little maintenance, and are usually smaller than other pumps of equal capacity (Figure 3.22). Centrifugal pumps can-

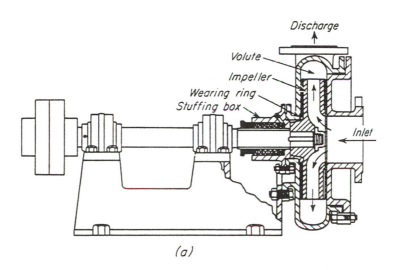

(a)

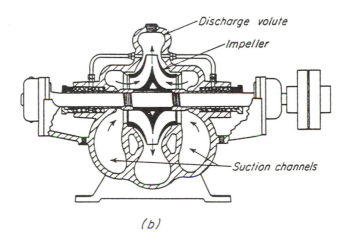

(b)

FIGURE 3.22. One of the most commonly used pumps for general, nonslurry fluid flow is the centrifugal pump. (a) single-suction and (b) double-suction. (Reproduced with permission from reference 8. Copyright 1985 McGraw-Hill.)

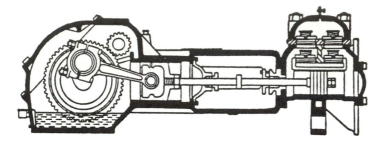

FIGURE 3.23.   A piston pump delivers high pressure and can handle viscous fluids. (Reproduced with permission from reference 11. Copyright 1964 Butterworth–Heinemann.)

not handle viscous liquids or slurries with high solids content, they do not develop high pressure, and they are not self priming.

High pressure can be handled by a *piston pump* (Figure 3.23), which can operate against considerable pressure without priming. A piston pump is a positive displacement pump.

A viscous fluid can also be successfully handled with a rotary pump such as a *gear pump* (Figure 3.24) or *Moyno pump* (Figure 3.25). The Moyno pump is especially useful for slurries because it uses a squeezing action to move the fluid.

Selecting the proper pump requires a calculation of the capacity needed and a knowledge of the process flow, the chemical's flow behavior, and the fluid's physical behavior.

We have now covered engineering problems for which we determined the pressure drop in the piping. In many situations, the pressure drop is fixed, and we must find the fluid's velocity or volumetric flow rate. Such problems can be solved by trial

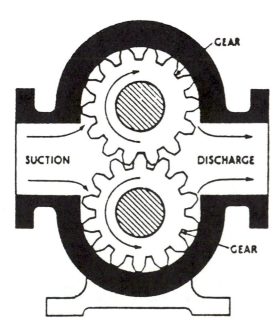

FIGURE 3.24.   A gear pump works well with viscous fluids. (Reproduced with permission from reference 11. Copyright 1964 Butterworth–Heinemann.)

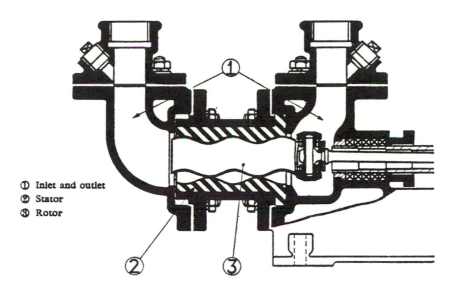

① Inlet and outlet
② Stator
③ Rotor

FIGURE 3.25.   A Moyno pump, a form of screw pump, moves fluids by turning a rotor within an elastic sleeve. (Reproduced with permission from reference 11. Copyright 1964 Butterworth–Heinemann.)

and error, assuming a velocity, then calculating the pressure drop; when the calculated and actual values match, the assumed velocity is the correct one.

There is a direct method of calculating the velocity, illustrated in Example 3.10. The technique is one not generally known to many practitioners.

EXAMPLE 3.10. USING A MODIFIED FRICTION FACTOR CHART

How many gallons of water at 20 °C can be delivered through a 400-m length of smooth pipe (0.15-m diameter) with a pressure difference of 1720 N/m²?

To solve this problem, we could use trial and error. We assume a flow rate, get f, and compute the pressure drop. When the calculated pressure drop matches the actual value, then flow rate is correct.

Another technique makes use of f (Figure 3.26). In this figure, f is plotted as a function of $\mathrm{Re}\sqrt{f}$ and

$$\mathrm{Re}\sqrt{f} = \frac{D\rho}{\mu}\sqrt{\frac{\Delta P D}{2L\rho}} \tag{3.40}$$

The value of the relationship is that $\overline{V}$ is eliminated. Hence, we can compute $\mathrm{Re}\sqrt{f}$ and then read the value of f.

$$\mathrm{Re}\sqrt{f} = \frac{(0.15\ \mathrm{m})\left(998\ \dfrac{\mathrm{kg}}{\mathrm{m}^3}\right)}{\left(0.001\ \dfrac{\mathrm{kg}}{\mathrm{m\text{-}s}}\right)}\left[\frac{\left(1720\ \dfrac{\mathrm{N}}{\mathrm{m}^2}\right)(0.15\ \mathrm{m})}{(400\ \mathrm{m})\left(998\ \dfrac{\mathrm{kg}}{\mathrm{m}^3}\right)(2)}\right]^{1/2}$$

$$\mathrm{Re}\sqrt{f} = 2695$$

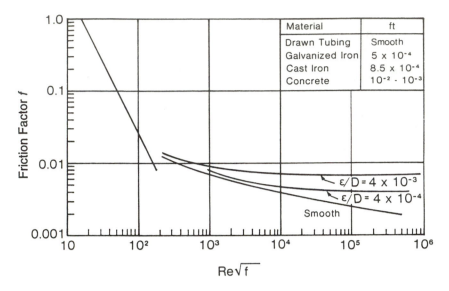

FIGURE 3.26. Calculating the velocity that is possible, given a pressure drop, is faster if one uses a modified friction factor plot, thus avoiding an iterative solution.

and

$$f = 0.0057$$

from Figure 3.26. Then,

$$\frac{\mathrm{Re}\sqrt{f}}{\sqrt{f}} = 2695/(0.0057)^{1/2} = 35,700$$

$$D\overline{V}\,\rho/\mu = 35,700$$

$$\overline{V} = (35,700)(0.001\ \mathrm{kg/m\text{-}s})/(0.15\ \mathrm{m})(998\ \mathrm{g/m^3})$$

$$\overline{V} = 0.238\ \mathrm{m/s}$$

$$Q = \pi D^2 \overline{V}/4 = \frac{\pi(0.15\ \mathrm{m})^2(0.238\ \mathrm{m/s})}{4} = 0.0042\ \mathrm{m^3/s}$$

and

$$Q = 67\ \mathrm{gallons/min}$$

The next example illustrates a problem that could arise at any time in a chemical plant. Our approach to the solution is to use judicious insights to overcome a lack of data and obtain a meaningful solution.

EXAMPLE 3.11. CALCULATING FLOW FROM A RUPTURED PIPE

A large, high-pressure chemical reactor contains water at an absolute pressure of $1.38 \times 10^7\ \mathrm{N/m^2}$ and a temperature of 20 °C. A 0.07-m inside diameter line connected to the reactor ruptures at a point 3 m from the reactor. What is the flow rate from the break?

Although flow from a burst pipe would be a nonsteady-state condition, we assume steady state at the time of the break to get the initial outflow. Starting with the Bernoulli equation

$$-F_h = \Delta\frac{P}{\rho} + \frac{\overline{V^2}}{2}$$

and taking surface 1 as the liquid surface in the tank and surface 2 at the break we obtain

$$\overline{V}_2 >>> \overline{V}_1$$

Notice that we have neglected the $\overline{V}_1^2/2$ term (because $\overline{V}_2 >>> \overline{V}_1$) and any potential energy effects (gZ) because kinetic energy will be very large. The $-dW_s/dm$ term for mechanical work is nonexistent.

The greatest frictional effects are due to two areas of the system, the sudden contraction from the tank to the pipe, called the *entrance effect*, and the piping itself.

$$F_h = K(\overline{V}_2^2/2) + 4f(L/D)(\overline{V}_2^2/2)$$

We include in the entrance effect only the contraction from tank to pipe because the expansion from the broken pipe takes place past boundary 2 (Figure 3.17). Substituting and solving for $\overline{V}_2$

$$\overline{V}_2 = [(-2\Delta P/\rho)/(1 + K + 4fL/D)]^{1/2}$$

From Figure 3.19, we find that K equals 0.5 because the line diameter is much smaller than the tank.

For the pipe, ϵ/D is 0.000043/0.07 or 0.00063, using an ϵ value for commercial pipe. Assuming Re to be large for such an ϵ/D, and noticing that, for such cases, the f versus Re relation is nearly flat (from Figure 3.15) we estimate that f is about 0.0043. Then,

$$\overline{V}_2 = \left\{ \frac{2(1.38 - 0.01) \times 10^7\ \text{N}}{\text{m}^2} \middle/ \left[\frac{998\ \text{kg}}{\text{m}^3}\left(1 + 0.5 + \frac{4\ (0.0043)(3\ \text{m})}{0.07\ \text{m}}\right)\right]\right\}^{1/2}$$

and

$$\overline{V}_2 = 110\ \text{m/s}$$

The Reynolds number would then be

$$\text{Re} = (0.07\ \text{m})(110\ \text{m/s})(998\ \text{kg/m}^3)/(0.001\ \text{kg/m-s}) = 7.7 \times 10^6$$

For this value of the Reynolds number, f is on the flat portion of the line, and the assumed f value is correct.

Then,

$$Q = (110 \text{ m/s})(\pi/4)(0.07 \text{ m})^2 = 0.423 \text{ m}^3/\text{s}$$

or

$$Q = 6800 \text{ gal./min}$$

FLOW AROUND SUBMERGED OBJECTS— CALCULATING A DRAG FORCE

CHARACTERIZING FLOW

Characterizing flow around a submerged object is more complicated than characterizing flow in a conduit. The basis for describing such flow is the drag force. Newton proposed that the drag force F on a sphere is

$$F = \pi r^2 \rho_{\text{air}} \, \overline{V}^2/2 \tag{3.41}$$

Correctly, the equation is

$$F/A = C_d \rho \, \overline{V}^2/2 \tag{3.42}$$

where C_d is the drag coefficient of an entity with the same role as f, and A is the object's surface area. C_d is a function of the particle Reynolds number Re_p,

$$\text{Re}_p = D_p \, \overline{V} \, \rho/\mu \tag{3.43}$$

and D_p is the particle diameter. Values for C_d corresponding to the Re_p are shown in Figure 3.27.

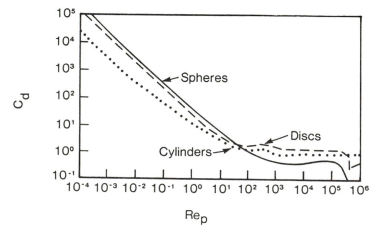

FIGURE 3.27.   To calculate the flow of a fluid around objects, we need to know the drag coefficient, which is shown here correlated with the Reynolds number Re_p for flow around particles. (Reproduced from reference 4. Copyright 1940 American Chemical Society.)

When a particle is very small,

$$C_d = 24/\text{Re}_p$$

and Stokes's law is obeyed. The particle settles slowly through the fluid. A balance between the buoyant and drag forces gives the velocity,

$$\overline{V} = 2r^2 g(\rho_{\text{particle}} - \rho_{\text{fluid}})/9\mu \tag{3.44}$$

The value of C_d is determined by the Re_p value,

$$C_d = b_1/\text{Re}_p{}^m \tag{3.45}$$

where b_1 and m have the values shown in Table 3.4 for spheres. A criterion for the proper range of fluid behavior can be expressed as

$$A = D_p[A_e\rho_{\text{fluid}}(\rho_{\text{particle}} - \rho_{\text{fluid}})/\mu^2]^{1/3} \tag{3.46}$$

where A_e is the acceleration of the particle. If Re_p is equal to 2.0 (the limit of Stokes's law), then A would be 3.3. The A value for the other two ranges would be ($3.3 < A < 43.6$) and ($43.6 < A < 2360$), respectively.

EXAMPLE 3.12. CALCULATING THE SETTLING RATE OF FINE DROPLETS

Drops of oil 15 μm in diameter are to be settled from an air mixture at 20 °C and 1 atm. Oil specific gravity is 0.90, and the settling time is to be 1 min. How high should the chamber be?

$$D_p = 15 \times 10^{-6} \text{ m}$$

$$\mu_{\text{air}} = 1.8 \times 10^{-5} \text{ kg/m-s}$$

$$\rho_{\text{particle}} = 898.6 \text{ kg/m}^3$$

$$A = (15 \times 10^{-6} \text{ m})\left[\frac{\left(9.8 \frac{\text{m}}{\text{s}^2}\right)\left(1.2 \frac{\text{kg}}{\text{m}^3}\right)\left(898.6 \frac{\text{kg}}{\text{m}^3}\right)}{\left(1.8 \times 10^{-5} \frac{\text{kg}}{\text{m-s}}\right)^2}\right]^{1/3}$$

TABLE 3.4. Coefficient Values for $C_d = b_1/\text{Re}_p{}^m$ for Spheres

Range	b_1	m
$\text{Re}_p < 2$	24	1.0
$2 < \text{Re}_p < 500$	18.5	0.6
$500 < \text{Re}_p < 200,000$	0.44	0.0

The density of air was neglected.

$$A = 0.479$$

This value is well below the limit of Stokes's law.

$$\overline{V} = 2r^2 g(\rho_{particle} - \rho_{fluid})/9\mu$$

$$\overline{V} = 2(7.5 \times 10^{-6} \text{ m})^2(9.8 \text{ m/s}^2)(898.6 \text{ kg/m}^3)/9(1.8 \times 10^{-5} \text{ kg/m-s})$$

$$\overline{V} = 0.0061 \text{ m/s}$$

In one minute the particles will settle $(0.0061 \text{ m/s})(60 \text{ s})$ or 0.366 m. The chamber should be 0.366 m tall.

NON-NEWTONIAN FLUID FLOW

Many fluids of industrial importance do not obey Newton's law of viscosity. *Shear thinning* or *pseudoplastic fluids,* for example, have viscosities that decrease with increasing shear stress. A *shear thickening* or *dilatant fluid* has a viscosity that increases with increasing shear rate. For a large number of non-Newtonian fluids, however, the shear stress is a function only of the shear rate, a relationship described by using an apparent viscosity μ_{app} in place of the Newtonian viscosity in the equation. So for non-Newtonian fluids, we can apply the relationship

$$\tau = -\mu_{app} dV/dy \tag{3.47}$$

$$\mu_{app} = \phi(dV/dy) \tag{3.48}$$

or use empirical equations to handle the flow calculations. One empirical equation is the power law

$$\tau = -K(dV/dy)^n \tag{3.49}$$

where K is the consistency index and n is the power law index. The values of K and n are found by plotting $\ln \tau$ against $\ln -dV/dy$.

A Reynolds number for non-Newtonian fluids that obey the power law has been correlated with the friction factor (Figure 3.28). The Reynolds number for the correlation is

$$Re' = D^{n'}\overline{V}^{(2-n')}\rho/g_c K' 8^{(n'-1)} \tag{3.50}$$

and

$$n \cong n' \tag{3.51}$$

Then,

$$K' = K[(3n + 1)/4n)]^n \tag{3.52}$$

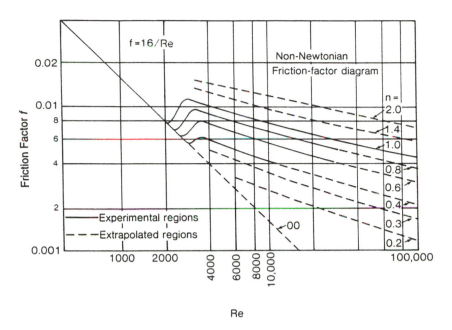

FIGURE 3.28. Non-Newtonian fluids that obey the power law have a friction factor correlated with the non-Newtonian Reynolds number (eq 3.50). (Reproduced with permission from reference 5. Copyright 1959 AIChE.)

The correlation is used in the same way as the Bernoulli equation is applied to Newtonian fluids. The extra data required are the power law constants n' and K'. These values are found by plotting $\ln \tau_{wall}$ (where τ_{wall} is equal to $R\Delta P/2L$) vs. $\ln 8\, \overline{V}/D$.

MIXING FLUIDS

Many chemical processes depend on fluid mixing for successful operation. When mixing a fluid, there are two main considerations: how much power and time agitation will require.

POWER REQUIREMENTS

The power is a function of several variables, including the speed of agitation N in revolutions per minute, the diameter D' of the impeller, and the fluid viscosity and density. Agitation can be compared among different systems and sizes by examining two related dimensionless groups, the power number P_o and the Reynolds number

$$P_o = Pg_c/N^3\rho(D')^5 \tag{3.53}$$

and

$$Re^* = N\rho(D')^2/\mu \tag{3.54}$$

P_o and Re have been correlated as shown in Table 3.5 and Figure 3.29 for agitation in a baffled tank. (The flow pattern in a baffled tank is depicted in Figure 3.30.) The

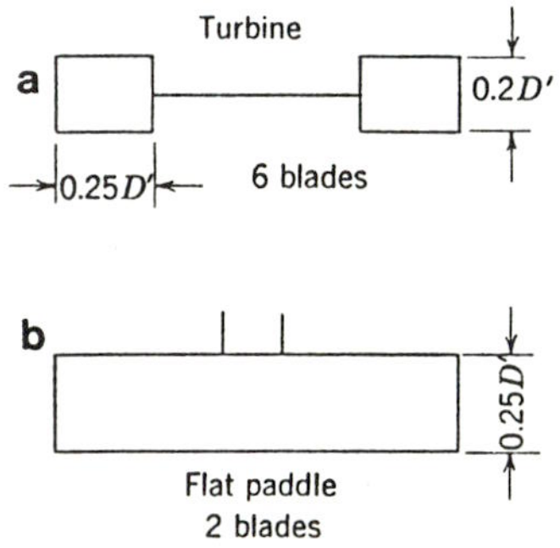

TABLE 3.5. Data for Baffled Tanks

Type of Impeller	Baffles (W/D)	Curve No. in Fig. 3.29
Part a	0.17	1
Part a	0.10	2
Part a	0.04	3
Same as Part a, two curved blades	0.10	4
Marine propeller, 3 blades, pitch = D'	0.10	5
Part b	0.10	6
Marine propeller, 3 blades, pitch = $2D'$	0.10	7

NOTE: For all types of impeller, D_t/D' is 3, z_l/D' is 2.7–3.9, z_i/D' is 0.75–1.3, and there are four baffles.

a Turbine

0.2D'

0.25D' 6 blades

b Flat paddle 2 blades

0.25D

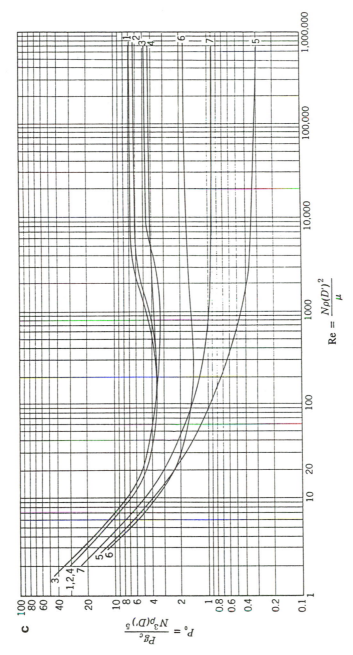

FIGURE 3.29. To use a power curve for a baffled tank, we need to know the pitch of the impeller; the impeller diameter D', the tank diameter D_t, the height of the liquid z_l, and the impeller height z_i. Opposite page, a and b are schematics. c, Power curves for baffled tanks. Pitch is the axial distance traced in one revolution when there is no slip between the agitator and the liquid. (Adapted from reference 6.)

curves in Figure 3.29 represent system geometries determined by the tank diameter D_t, liquid height Z_l, impeller or agitator elevation Z_i, and baffle width W. A similar correlation for P_o and Re for unbaffled tanks has been developed (Figure 3.31 and Table 3.6). All of the curves in Figure 3.31 show the laminar–transition turbulent behavior found earlier for flow in ducts, around objects, and for non-Newtonian fluids. Each of the curves is specific for a given agitator and system geometry.

The foregoing provides information only about the amount of power supplied to a liquid under given conditions with a certain impeller. It says nothing about the nature of the actual blending or the amount of power needed to suspend solid particles in a liquid. For these processes, we turn to empirical equations, eqs 3.55 and 3.56 to calculate the power needed to suspend a solid.

$$P g_c / g \rho_m V_m \overline{V} = (1 - \epsilon_m)(D_t / D_i)^{1/2} e^{4.35\beta} \tag{3.55}$$

and

$$\beta = [(Z_s - E)/D_t] - 0.1 \tag{3.56}$$

where ρ_m is the density, and V_m is the volume of the solid–liquid suspension, not including the clear zone above Z_s (the top of the suspension), $\overline{V}$ is the terminal settling velocity calculated from Stokes's law, ϵ_m is the volume fraction of liquid in the suspension region, and E is the clearance between the impeller and the tank bottom.

For blending, little data exist except for a correlation of blending times for miscible liquids using turbine agitators (Figure 3.32). The ordinate is a complicated function, f_t, defined as

$$f_t = \frac{t_t N^{2/3} D'^{4/3} g^{1/6} D'^{1/2}}{Z_l^{1/2} D_t^{3/2}} \tag{3.57}$$

where t_t is the blending time, N is agitation speed, D' is the agitator diameter, Z_l is the height of the liquid, and D_t is the tank diameter.

An example of an agitation calculation follows for an unbaffled tank system. In this problem, we see the effect of the vortex on the power requirement for the sys-

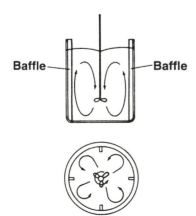

FIGURE 3.30. Baffles and impellers in a tank combine to create circulation in both the vertical and horizontal directions, respectively, as well as to damp the vortex in the center. (Reproduced with permission from reference 6. Copyright 1950 AIChE.)

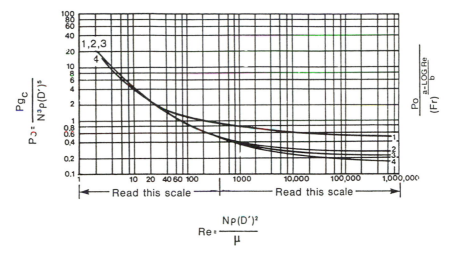

FIGURE 3.31. To calculate the power requirement for mixing in an unbaffled tank, we need to know the same information as for Figure 3.29. Power curves for unbaffled tanks. (Reproduced with permission from reference 6. Copyright 1950 AIChE.)

tem. The example also illustrates the importance of geometric similarity (ratios of parameters such as tank diameter to agitator diameter) in such calculations.

EXAMPLE 3.13. CALCULATING THE POWER NEEDED TO AGITATE A LIQUID

Calculate the power required for agitation with a 3-blade marine impeller of 0.6-m pitch and 0.6-m diameter operating at 100 rpm in an unbaffled tank containing water at 20 °C. The tank diameter is 1.8 m, depth of liquid is 1.8 m, and the impeller is located 0.6 m from the tank bottom. We see that $\mu = 0.001$ kg/m-s, $\rho = 998$

TABLE 3.6. Data for Unbaffled Tanks

Type of Impeller	D_t/D'	a	Curve No. in Fig. 3.31
Marine propellers, 3 blades, pitch = $2D'$	3.3	1.7	1
Marine propellers, 3 blades, pitch = $1.05D'$	2.7	2.3	2
Marine propellers, 3 blades, pitch = $1.04D'$	4.5	0	4
Marine propellers, 3 blades, pitch = D'	3	2.1	3

NOTE: For all types of impeller, z_l/D' is 2.7–3.9, z_i/D' is 0.75–1.3, and b is 18.

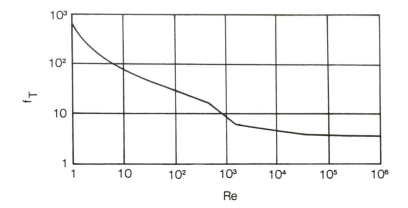

FIGURE 3.32.   The blending time function f_t is a complicated function; other blending correlations are even more empirical. (Reproduced with permission from reference 7. Copyright 1960 AIChE.)

kg/m³, $D_t/D' = 3.0$, $Z_l/D' = 3.0$, and $Z_i/D' = 1.0$. For the data given, the proper curve (Figure 3.31) is number 4. Hence,

$$Re = D'^2 N/\mu = (0.6 \text{ m})^2(100/60 \text{ s})(998 \text{ kg/m}^3)/0.001 \text{ kg/m-s}$$

$$Re = 5.99 \times 10^5$$

Because Re > 300, we must read the right-hand scale of Figure 3.31, $P_o/(Fr)^x$, where x is $(a - \log Re)/b$, a and b are constants, and Fr is the Froude number (N^2D'/g). Then, from Table 3.6 (Figure 3.31), the values of a and b are, respectively, 2.1 and 18, and, from the chart, $P_o/(Fr)^x$ equals 0.23.

$$x = (a - \log Re)/b = (2.1 - 5.82)/18 = -0.206$$

$$P_o/Fr^{-0.206} = 0.23$$

$$Fr = (100/60 \text{ s})^2(0.6 \text{ m})/9.81 \text{ m/s}^2) = 0.170$$

and

$$P_o = 0.23/0.173^{0.206} = 0.33$$

Then,

$$P = P_o N^3 D'^5 \rho = (0.33)(100/60 \text{ s})^3(998 \text{ kg/m}^3)(0.6 \text{ m})^5 = 118.6 \text{ watts}$$

$$P = (118.6 \text{ watts})(HP/746 \text{ watts}) = 0.16 \text{ } HP$$

FLOW THROUGH PACKED BEDS, POROSITY, AND SPHERICITY

Packed beds are used in a wide array of process equipment, including absorption columns and fixed-bed chemical reactors. Fluid flow through a packed bed is a complex process that we can visualize as flow through a collection of intermeshed tubes of varying cross section. This collection creates a tortuous path for the fluid, which enhances heat and mass transfer. We take a semiempirical approach to calculations for these flows because of their complexity.

The approach uses a hydraulic radius to derive appropriate flow equations. For a packed bed,

$$R_h = \text{volume available for flow/total wetted surface} \qquad (3.58)$$

$$R_h = \text{(volume of voids/bed volume)/(wetted surface/bed volume)} = \epsilon/a \quad (3.59)$$

where ϵ is porosity, and a equals $a_V(1 - \epsilon)$ where a_V is the total particle surface divided by the volume of the particles.

Also useful in packed-bed calculations is the superficial velocity $\bar{V}_o$ (defined as the approach velocity to the bed, or the velocity without the packing present), the packed bed height L, and particle diameter D_p. The Reynolds number for a packed bed is

$$\text{Re} = \frac{D_p \bar{V}_o \rho}{\mu} \qquad (3.60)$$

For both laminar and turbulent flow through a bed,

$$\frac{\Delta P g_c}{L} \frac{\phi_s D_p}{\rho \bar{V}_o} \frac{\epsilon^3}{1 - \epsilon} = \frac{150(1 - \epsilon)}{\phi_s D_p \bar{V}_o \dfrac{\rho}{\mu}} + 1.75 \qquad (3.61)$$

which is the Ergun equation. The term ϕ_s is the sphericity and is obtained by the relation

$$\phi_s = (6/D_p)(\text{particle surface area/particle volume}) \qquad (3.62)$$

Some typical sphericities for various packings are given in Table 3.7.

For the laminar range (Re < 1.0), the Ergun equation (eq 3.61) is

$$\frac{\Delta P g_c}{L} \frac{\phi_s^2 D_p^2}{\bar{V}_o \mu} \frac{\epsilon^3}{(1 - \epsilon)^2} = 150 \qquad (3.63)$$

For turbulent flow (Re > 1000),

$$\frac{\Delta P g_c}{\rho L} \frac{\phi_s D_p}{\bar{V}_o^2} \frac{\epsilon^3}{(1 - \epsilon)} = 1.75 \qquad (3.64)$$

For calculations in the transition range (1.0 < Re < 1000), use eq 3.61.

TABLE 3.7.   Sphericity (ϕ_s) Values for Packing
Materials

Material	ϕ_s
Sphere, cube, or cylinder ($L = D$)	1.0

Raschig ring,
  inside diameter L
   $= 1/2$ outside diameter D
	0.58

Raschig ring,
  inside diameter $L =$
  $3/4$ outside diameter D
	0.33

Berl saddles	0.3
Sharp-pointed sand particles	0.95
Rounded sand	0.83
Coal dust	0.73
Crushed glass	0.65
Mica flakes	0.28

For cases when the packed bed consists of a mixture of different particle sizes, a surface mean diameter $\overline{D}_s$ must be used. This diameter is obtained by using the following equations and the total number of each size particle N_i or the mass fraction of each particle size x_i:

$$\overline{D}_s = \sum_{i=1}^{n} N_i D_{pi}^{3} / \sum_{i=1}^{n} N_i D_{pi}^{2} \tag{3.65}$$

and

$$\overline{D}_s = 1 / \sum_{i=1}^{n} x_i / D_{pi} \tag{3.66}$$

REFERENCES

1. *Chemical Engineers' Handbook*, 5th ed.; Perry, R. H.; Chilton, C. H., Eds.; McGraw-Hill: New York, 1973, out of print; 7th ed. in press.
2. Moody, L. W. *Trans. ASME* **1944,** *66,* 672.
3. Crane Technical Paper No. 410, Crane Company, Chicago, IL. This is actually a spiral-bound book called *Flow of Fluids Through Valves, Fittings, and Pipe*, which contains all sorts of practical engineering data and examples of flow problems for the hands-on plant engineer.
4. Lapple, C. E.; Shepherd, C. B. *Ind. Eng. Chem.* **1940,** *32,* 605.
5. Dodge, D. W.; Metzner, A. B. *AIChE J.* **1959,** *5,* 189.
6. Rushton, J. H.; Costich, E.; Everett, H. J. *Chem. Eng. Progr.* **1950,** *46,* 395, 467.
7. Norwood, K. W.; Metzner, A. B. *AIChE J.* **1960,** *6,* 432.
8. McCabe, W. L.; Smith, J. C.; Harriott, P. *Unit Operations of Chemical Engineering;* McGraw-Hill: New York, 1985.
9. DeNevers, N. *Fluid Mechanics;* Addison-Wesley: Reading, MA, 1970.
10. Bird, R. B.; Stewart, W. E.; Lightfoot, E. N. *Transport Phenomena;* John Wiley and Sons: New York, 1960.

11. Coulson, J. M.; Richardson, J. F. *Chemical Engineering;* Pergamon: London, 1978; Vols. 1 and 2.
12. *Fluid Meters, Their Theory and Application;* ASME International: New York, 1959.
13. Tuve, G. L.; Sprenkle, R. E. *Instruments* **1933,** *6*, 201.
14. Rohsenow, W. M.; Choi, H. *Heat, Mass, and Momentum Transfer;* Prentice-Hall: Englewood Cliffs, NJ, 1961.

FURTHER READING

Bennett, C. O.; Myers, J. E. *Momentum, Heat, and Mass Transfer*, 3rd ed.; McGraw-Hill: New York, 1982.

Fahien, R. *Fundamentals of Transport Phenomena;* McGraw-Hill: New York, 1983.

Foust, A. S.; Wenzel, L. A.; Clump, C. W.; Maus, L.; Anderson, L. *Principles of Unit Operations*, 2nd ed.; reprint of 1980 ed.; Krieger, 1990.

Geankoplis, C. J. *Transport Processes and Unit Operations*, 2nd. ed.; Allyn and Bacon: Boston, 1989.

Greenkorn, R. A.; Kessler, D. P. *Transfer Operations;* McGraw-Hill: New York, 1972, out of print.

John, J. E.; Haberman, W. L. *Introduction to Fluid Mechanics*, 3rd ed.; Prentice-Hall: Englewood Cliffs, NJ, 1988.

Knudsen, J. D.; Katz, D. L. *Fluid Dynamics and Heat Transfer;* McGraw-Hill: New York, 1958, out of print.

Prasuhn, A. L. *Fundamentals of Fluid Mechanics;* Prentice-Hall: Englewood Cliffs, NJ, 1980, out of print.

4

Heat Transfer

From the first law of thermodynamics in Chapter 2,

$$\Delta U = Q - W \tag{4.1}$$

where U is internal energy, Q is heat, and W is work. For heat transfer, there must be a temperature difference, ΔT, which is a driving force. Heat can be transferred by different mechanisms: conduction, convection, and radiation.

CONDUCTION

In any system, energy is transferred from a high-temperature to a low-temperature region. We will calculate heat transfer as a *flux*, which is power (heat per unit time) transferred per unit area in each of the three coordinate directions. The equation governing this transfer is Fourier's law, which in each dimension is

$$q_x' = -kdT/dx \tag{4.2}$$

$$q_y' = -kdT/dy \tag{4.3}$$

and

$$q_z' = -kdT/dz \tag{4.4}$$

Because heat flux is a vector,

$$q' = k[(\partial T/\partial x)\underline{i} + (\partial T/\partial y)\underline{j} + (\partial T/\partial z)\underline{k}] \tag{4.5}$$

where $\underline{i}$, $\underline{j}$, and $\underline{k}$ are unit vectors in the x, y, and z direction, q' is the energy or heat flux in kJ/s-m^2, k is the thermal conductivity in kJ/s-m-K, and T is temperature in K.

If the material has an internal heat source and if the heat transfer is unsteady (i.e., it varies with temperature), then the equation of energy is

$$\partial/\partial x(k\partial T/\partial x) + \partial/\partial y(k\partial T/\partial y) + \partial/\partial z(k\partial T/\partial z) + \dot{q} = \rho C_p\, \partial T/\partial t \quad (4.6)$$

where $\dot{q}$ is a heat source in units of energy generated per unit volume and unit time, ρ is density, C_p is the specific heat, and t is time.

CONVECTION—HEAT TRANSFER WITH MOTION

The second mode of heat transfer, convection, involves at least one component in motion, for example, a liquid flowing past a solid or a gas and liquid both flowing. Convection is conduction together with motion. Because this is a complex system, we use the empirical expression

$$q = hA\Delta T \quad (4.7)$$

where h is a heat transfer coefficient, A is the area, and ΔT is the temperature difference. In some instances h can be determined by calculation, but it is usually determined experimentally.

RADIATION—HEAT TRANSFER THROUGH A VACUUM

The third mode of heat transfer is radiation. In radiation, heat can even be transferred through a vacuum, unlike conduction and convection, where heat is transferred through a material. The mechanism for radiation is the propagation of electromagnetic radiation due to a temperature difference.

The governing equation for radiation is the Stefan–Boltzmann law

$$q = \sigma A(T_1^4 - T_2^4) \quad (4.8)$$

where σ is a proportionality constant (the Stefan–Boltzmann constant), A is the area, and T_1 and T_2 are the temperatures of the source and sink for the radiation, respectively.

STEADY-STATE HEAT CONDUCTION—q STEADY WITH TIME

To understand the concept of steady-state conduction, for which heat transfer remains constant with time t, we can carry out a simple experiment. The experiment consists of taking a slab of material (originally at temperature T_1) and suddenly raising one face to temperature T_2 ($T_2 > T_1$). The net result will be a flow of energy from the area of higher temperature to the area of lower temperature. In addition, a temperature profile will develop in the cross section of the slab (*see* the curved lines of Figure 4.1a) until a state is reached when the profile changes no further. When this occurs, the profile will be linear.

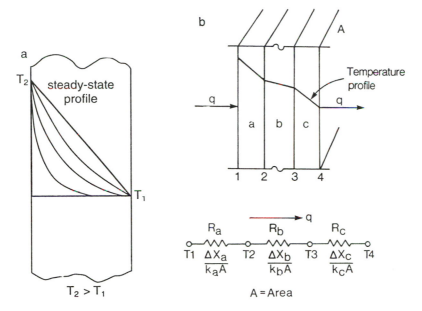

FIGURE 4.1. (a) If one face of a solid slab of material at temperature T_1 is subjected to a higher temperature T_2, the material will conduct heat. The temperature gradient across the slab will change (see the curved lines) until, at some point, the profile is linear and stable with time. (b) If layers of different materials are placed in series, each will have a distinct linear temperature profile, defined by thermal conductivity k_a, k_b, k_c, etc. At steady-state conditions, the rate of heat transferred, q, is the same through all layers. Individual material thickness and the material's thermal conductivity determine the individual resistances to heat transfer, much like electrical resistances. (Reproduced with permission from reference 22. Copyright 1981 McGraw-Hill.)

Now, if we consider the conduction at steady state in only one dimension,

$$q = -kA \, dT/dx \qquad (4.9)$$

Integrated (holding k, A, and q constant), eq 4.9 becomes

$$q = kA(T_1 - T_2)/\Delta x \qquad (4.10)$$

This equation describes the linear change in temperature across the slab for a given material. If a number of materials are put together as shown in Figure 4.1b, then eq 4.10 can be written for each material and the heat, q, is conducted through all of the materials where

$$q = k_a A(T_1 - T_2)/\Delta x_a = k_b A(T_2 - T_3)/\Delta x_b = k_c A(T_3 - T_4)/\Delta x_c \qquad (4.11)$$

To find the change in energy through the composite related to the end point temperatures, we add the terms for each coordinate to get

$$q = (T_1 - T_4)/(\Delta x_a/k_a A + \Delta x_b/k_b A + \Delta x_c/k_c A) \qquad (4.12)$$

Equation 4.12 is analogous to Ohm's law for electrical conduction, where the term $\Delta x/kA$ is the resistance to heat transfer.

If we consider one-dimensional radial conduction in a cylinder instead of linear conduction through a flat slab, we see that

$$q = -kA \, dT/dr \tag{4.13}$$

Because the area A equals $2\pi rL$,

$$dr/r = d \ln r = (-k2\pi rL/q) \, dT \tag{4.14}$$

Solving eq 4.14 for a hollow cylinder with an inner radius r_i and outer radius r_o, yields

$$q = 2\pi kL(T_i - T_o)/\ln (r_o/r_i) \tag{4.15}$$

If concentric sections of different materials make up the cylinder, then

$$q = 2\pi L\Delta T_{\text{overall}}/[(\ln r_2/r_1)/k_a + (\ln r_3/r_2)/k_b + (\ln r_4/r_3)/k_c] \tag{4.16}$$

where $(\ln r_o/r_i)/k$ is the resistance to heat transfer for each material.

We shall now work a few examples involving steady-state heat conduction with multiple resistances in both slabs and cylinders. The heat flow experienced by all parts of the system is the same (analogous to electric current), which makes it possible to determine a given temperature at any point in the system.

EXAMPLE 4.1. HEAT TRANSFER THROUGH A COMPOSITE, FLAT SURFACE

A cold storage room has walls constructed of 0.102-m-thick corkboard sandwiched between two wooden walls each 0.0127 m thick. What is the rate of heat loss if the inside wall surface temperature is -12.2 °C and the outside wall surface is 21.1 °C? Also, what is the temperature at the interface between the outer wall and the corkboard? Given k_{cork} equals 0.0415 W/m-°C, and $k_{\text{wooden wall}}$ equals 0.1073 W/m-°C,

$$q = \Delta T_{\text{overall}}/[\Delta x_a/k_a A + \Delta x_b/k_b A + \Delta x_c/k_c A]$$

$$q = (21.1 + 12.2)/(1/m^2)[0.0127/0.1073 + 0.102/0.0415 + 0.0127/0.1073]$$

and

$$q = 11.51 \text{ W}$$

Then

$$q/A = -k\Delta T/\Delta x$$

$$q/A = 11.51 \text{ W/m}^2 = (0.1073)(21.1 - T_{\text{interface}})/0.0127$$

and

$$T_{\text{interface}} = 19.7 \text{ °C}$$

EXAMPLE 4.2. HEAT TRANSFER THROUGH A CYLINDER

A steel pipe of outside diameter 0.051 m is insulated with a 0.0064-m-thick layer of asbestos followed by a 0.0254-m-thick layer of fiberglass. If the pipe wall temperature is 315.6 °C and the outside insulation is 37.8 °C, what is the temperature between the asbestos and the fiberglass? Given $k_{asbestos}$ equal to 0.166 W/m-°C and $k_{fiberglass}$ equal to 0.0485 W/m-°C,

$$q = 2\pi L \Delta T_{overall}/[(\ln r_2/r_1)/k_a + (\ln r_3/r_2)/k_b]$$

$$q = 2\pi L(315.6 - 37.8)/$$
$$[(\ln 0.0318/0.0254)/0.166] + [(\ln 0.057/0.0318)/0.0485]$$

$$q/L = 129.5 \text{ W/m}$$

and

$$q/L = 129.5 \text{ W/m} = 2\pi(315.6 - T_{interface})/(\ln 1.25/0.166)$$

$$T_{interface} = 288.3 \text{ °C}$$

STEADY-STATE HEAT CONDUCTION WITH A HEAT SOURCE—HEATED WIRES, CHEMICAL REACTORS, AND PHASE CHANGES

A SOURCE OF HEAT

In many processes, the material has a source of heat $\dot{q}$ as in eq 4.6. The source can be a chemical reaction (nuclear reactions included), a phase change, or electrical heating.

To evaluate the effect of a heat source, we can consider a plane wall with a heat source uniformly distributed through the slab. We assume steady-state conditions ($\partial T/\partial t$ is zero) and no conduction in the y or z direction. Equation 4.6 holds, and

$$\partial/\partial x(k\partial T/\partial x) + \dot{q} = 0 \tag{4.17}$$

If k is assumed independent of position, then

$$\partial^2 T/\partial x^2 + \dot{q}/k = 0 \tag{4.18}$$

This is the basic differential equation, which describes one-dimensional conduction with internal heat generation.

For cylindrical coordinates, the overall conduction equation is

$$\partial^2 T/\partial r^2 + (1/r)(\partial T/\partial r) + (1/r^2)(\partial^2 T/\partial \theta^2) + \partial^2 T/\partial z^2 + \dot{q}/k = (\rho C_p/k)(\partial T/\partial t) \tag{4.19}$$

If we assume no conduction in the θ or z direction and steady-state conditions,

$$\partial^2 T/\partial r^2 + (1/r)(\partial T/\partial r) + \dot{q}/k = 0 \tag{4.20}$$

which is the differential equation for one-dimensional (r direction) conduction with heat generation for cylinders.

Solution of eq 4.20 for the case of electrical heating by a wire is given in Example 4.3. A similar approach can be used for other heat generation.

EXAMPLE 4.3. HEAT TRANSFER FROM A WIRE

A 0.0032-m-diameter, 0.3048-m-long wire carries 10 V. The outer surface temperature is maintained at 93.3 °C. What is the temperature at the center of the wire? (k_{wire} is 22.5 W/m-°C; resistivity is 70 $\mu\Omega$-cm.)

$$\dot{q}\pi r^2 L = \text{voltage}^2/\text{resistance}$$

$$\dot{q} = \text{voltage}^2/\pi r^2 LR = 10^2/R\pi(0.16 \text{ cm})^2(30.48 \text{ cm})$$

$$R' = (70 \ \mu\Omega\text{-cm})(30.48 \text{ cm})/\pi(0.16 \text{ cm})^2$$

$$R' = 0.0268 \ \Omega$$

and

$$\dot{q} = 1.539 \times 10^9 \text{ W/m}^3$$

Then solving

$$\partial^2 T/\partial r^2 + 1/r(\partial T/\partial r) + \dot{q}/k = 0$$

for

$$r = 0 \text{ and } T = T_o$$

to

$$r = R \text{ and } T = T_w$$

gives

$$T_o = (1.539 \times 10^9 \text{ W/m}^3)(0.0016 \text{ m})^2/4(22.5 \text{ W/m-°C}) + 93.3 \text{ °C}$$

$$T_o = 137.1 \text{ °C}$$

UNSTEADY-STATE HEAT CONDUCTION

CHANGES WITH TIME

In many processes, the temperature of the material changes with time, and the temperature difference that drives the heat transfer cannot be considered steady with time. We might, for example, need to know how long it will take for a material, such as an injection-molded plastic, to cool. Hence, examining eq 4.6, but assuming no source of heat in the material

$$\partial/\partial x(k\partial T/\partial x) + \partial/\partial y(k\partial T/\partial y) + \partial/\partial z(k\partial T/\partial z) = \rho C_p \ \partial T/\partial t \qquad (4.21)$$

Finding solutions to eq 4.21 for various heat transfer systems is often investigated because, although the equation is complicated, it can be solved using analytical techniques. Entire books exist (*1, 2*) that tabulate solutions to heat transfer to or from specific geometries. Additionally, charts have been developed (*3*) that present solutions to eq 4.21 in graphical form for various configurations. A typical chart for unsteady-state conduction in a large flat slab is called the Gurney–Lurie chart (Figure 4.2).

To explain Figure 4.2, we must define some terms. Y is a reduced temperature $(T_1 - T)/(T_1 - T_0)$, where T_1 is surface temperature, T is the temperature at a given point at a given time, and T_0 is the temperature at the given point at zero time. X is a dimensionless time and is equal to $kt/\rho C_p x_1^2$ where x_1 is half of the slab thickness. The term n is a dimensionless position x/x_1, and m is the ratio of k to h (the film or heat transfer coefficient) multiplied by x_1.

Example 4.4 uses the Gurney–Lurie chart to delineate quickly an unsteady-state temperature in a slab. (An analytical calculation of such a temperature involves effort and differential equations.) Graphical solutions also exist for spheres and cylinders. There are a number of practical applications of the solution of Example 4.4, such as the cooling of a metal billet or the heating of a slab.

EXAMPLE 4.4. UNSTEADY-STATE HEATING OF A SLAB

How much time would be needed to raise the centerline temperature of a hard rubber slab to 132.2 °C if it were sandwiched between two electrically heated metal plates at 141.7 °C? The rubber slab is originally at 26.7 °C and is 0.0127 m thick. The thermal diffusivity ($k/\rho C_p$) of the slab is 7.56×10^{-8} m^2/s. Assume h from the metal to the rubber is 5678 W/m^2-°C.

First we must obtain the parameters needed for Figure 4.2.

$$n = x/x_1 = 0/0.0064 = 0$$

$$m = (0.159 \text{ W/m-°C})/(5678 \text{ W/m}^2\text{-°C})(0.0064)$$

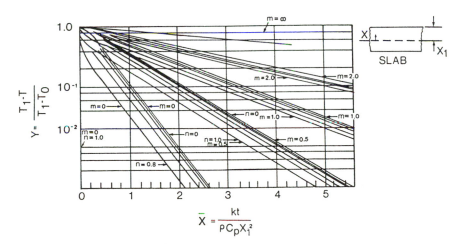

FIGURE 4.2.   Unsteady-state heat conduction is commonplace in industrial processes. Many graphical solutions, such as this chart for conduction through slab, have been developed. Y is a reduced temperature, and X represents time in dimensionless form. (To see how to use this chart, refer to Example 4.4.) (Reproduced from reference 3. Copyright 1923 American Chemical Society.)

and

$$m = 0.00438$$

The range of m is zero to one. Because the value of m just found is very small, we set m equal to zero. Then

$$Y = (141.7 - 132.2)/(141.7 - 26.7) = 0.0826$$

In Figure 4.2, we can find Y equal to 0.0826 and find the intersection with the line for $n = 0$ and $m = 0$. At this point,

$$\overline{X} = 1.13 = kt/\rho C_p x_1{}^2$$

Then solving for t, we find that the time required is 612 s.

CONVECTIVE HEAT TRANSFER—FORCED AND FREE

When heat transfer is accompanied by fluid motion, the heat transfer mechanism is convection. The semiempirical equation governing this mode of heat transfer is

$$q = hA\Delta T$$

Convection can be forced or free. Forced convection involves flow patterns determined by an external force, whereas free convection is created by the buoyancy of a heated fluid. (A typical case of free convection is the heat wave seen rising from a radiator or a hot asphalt pavement.) The difference between forced and free convection is illustrated in Figure 4.3. In forced convection, the movement of cold air is assisted by a fan, whereas in free convection buoyancy forces alone provide cold air.

Table 4.1 lists typical values for the heat transfer coefficient h. They represent ranges, not actual values of h, and they should not be used to calculate the heat transfer q, but to judge effects due to the types of fluids, effects of phase changes, etc., and their impact on convection. Most of the time, a value for h is determined experimentally and empirically.

FIGURE 4.3.   Convection can be forced (fan-assisted) or free.

TABLE 4.1   Value Ranges for the Convective Heat
Transfer Coefficient h

Situation	Material	$h(\text{W/m}^2\text{-}°\text{C})$
Free convection	gases	5.7–22.7
	liquids	113.6–681.4
	boiling water	113.6–2,271
Forced convection	gases	11.4–113.6
	viscous fluids	56.8–11,360
	water	568–11,360
Condensing vapors		1,136–113,600

Table 4.1 illustrates features of heat transfer. For example, phase changes (boiling water and condensing vapors) have high heat transfer coefficients. For both free and forced convection, heat transfer coefficients are lower for gases than for liquids. Gases have smaller coefficients than liquids because gases are poor conductors of heat and act as insulators. The purpose of foamed insulation, for example, is to trap air and improve the insulation quality.

DIMENSIONLESS NUMBER FOR COMPARING CONDUCTION AND CONVECTION—THE NUSSELT NUMBER

Heat transfer coefficients relate to the dimensionless groups mentioned earlier. Because flow is involved, the Reynolds number plays a role. Additionally, there is a group relating to heat transfer itself. This is the Prandtl number, Pr, defined as

$$\text{Pr} = \text{momentum diffusivity/thermal diffusivity} \qquad (4.22)$$

$$\text{Pr} = (\mu/\rho)/(k/\rho C_p) = \frac{C_p\mu}{k}$$

When the Prandtl and Reynolds numbers are multiplied

$$\text{PrRe} = \frac{\text{heat transport by convection}}{\text{heat transport by conduction}}$$

and

$$\text{PrRe} = \frac{\dfrac{\rho C_p V(T_0 - T_1)}{D}}{\dfrac{k(T_0 - T_1)}{D^2}} \qquad (4.23)$$

The correlation of h then involves a new dimensionless group, the Nusselt number and the function

$$\text{Nusselt number} = hD/k = \phi(\text{Re, Pr, } L/D, \mu_b/\mu_w) \qquad (4.24)$$

The other terms in the function besides Pr and Re are for entrance effects (L/D) or laminar flow and the change in the ratio of fluid viscosity in the bulk fluid and at the wall (μ_b/μ_w), which is sensitive to changes in temperature.

In natural convection, the Nusselt number takes a different form because of the influence of buoyancy forces on the system.

$$\text{Nusselt number} = hD/k = \phi(\text{Gr, Pr}) \qquad (4.25)$$

where Gr is the Grashof number ($L^3\rho^2 g\beta\Delta T/\mu^2$), in which L is a characteristic length, β is the coefficient of volume expansion, and ΔT is the temperature difference between the wall and fluid. When the Grashof number is compared to the Reynolds number,

$$\frac{\text{Grashof number}}{(\text{Reynolds number})^2} = \frac{\text{Gr}}{\text{Re}^2} = \frac{\rho\beta g(T_0 - T_1)}{\rho V^2/D} = \frac{\text{buoyancy forces}}{\text{inertial forces}} \qquad (4.26)$$

There are cases where a convective heat transfer problem can be solved analytically (for example, for steady-state heat conduction in laminar flow of a viscous fluid). In most cases, h is determined experimentally and empirically as we shall see in the discussion of empirically derived equations for h.

THE OVERALL HEAT TRANSFER COEFFICIENT, U—COMBINING CONDUCTIVE AND CONVECTIVE HEAT TRANSFER EFFECTS

Heat transfer frequently combines conduction and convection. To determine the heat transfer in such cases, we use the concept that heat transfer through the entire system equals that through any segment (assuming steady-state heat transfer), a method we applied when calculating conduction through multiple resistances. Heat transferred q is related to the overall system temperature difference ΔT and the surface area A. This transfer leads to an overall heat transfer coefficient U that combines both conduction and convection. (U is not internal energy as discussed in thermodynamics in Chapter 2.)

To understand U, we consider the system (Figure 4.4) for which we earlier discussed heat transfer by conduction through a series of adjacent materials. This treatment can be combined with convective heat transfer to give an overall heat transfer coefficient. On an overall basis, the heat transfer is

$$q = AU(T_i - T_o) \qquad (4.27)$$

Because the heat transfer by convection and conduction are equal,

$$q = h_o 2\pi r_3 L(T_3 - T_o) \qquad (4.28)$$

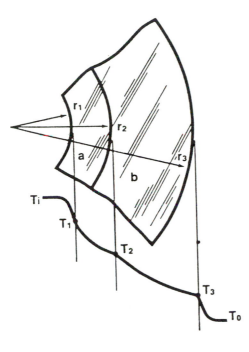

FIGURE 4.4. The overall heat transfer coefficient U for a composite cylinder accounts for resistances to conduction from the layered materials, a and b, and to convection at the inner and outer radii, r_1 and r_3. (Reproduced with permission from reference 23. Copyright 1961 Prentice-Hall.)

and

$$q = h_i 2\pi r_1 L(T_i - T_1)$$

where h_o and h_i are the coefficients for the outside and inside of the cylinder, and T_o and T_i are the temperature of the bulk fluid outside and inside the cylinder. Combining both conduction and convection in the cylinder,

$$1/AU = 1/2\pi Lr_1 h_i + \ln (r_2/r_1)/2\pi Lk_a + \ln (r_3/r_2)/2\pi Lk_b + 1/2\pi Lr_3 h_0 \quad (4.29)$$

If U is defined for an area A_1 equal to $2\pi r_1 L$, then,

$$U_1 = 1/[1/h_i + (r_1/k_a) \ln (r_2/r_1) + (r_1/k_b) \ln (r_3/r_2) + r_1/r_3 h_0] \quad (4.30)$$

This definition of the overall heat transfer coefficient could also be based on A_2 or A_3 because $A_1 U_1$ equals $A_2 U_2$ equals $A_3 U_3$ (Table 4.2).

DESIGN EQUATIONS FOR CONVECTIVE HEAT TRANSFER

Appropriate design equations for convective heat transfer coefficients h are found from empirical relationships between the Nusselt number (hD/k) and other dimensionless groups derived from experimental data.

For laminar flow in tubes, the average h value (4) can be derived from

$$hD/k = 1.86(RePr)^{0.33}(D/L)^{0.33}(\mu_b/\mu_w)^{0.14} \quad (4.31)$$

To use eq 4.31, all of the fluid properties are evaluated at the mean bulk temperatures of the fluids at the inlet and the outlet of the tubes.

TABLE 4.2 Overall Heat Transfer Coefficient U
for Heat Transfer Equipment

Type of Equipment	$U(\text{W/m}^2\text{-}°\text{C})$
Stabilizer reflux condenser	533.8
Oil preheater	613.3
Reboiler (condensing steam to boil milk)	2839.2–4542.6
Air heater (molten salt to air)	34.1
Steam-jacketed vessel evaporating milk	2839.2

When the tube is unusually long (i.e., D/L approaches zero), h is determined by another relationship (5). When D/L equals zero, the Nusselt number is 3.66.

$$hD/k = 3.66 + [0.0668(D/L)\text{RePr}]/[1 + 0.04(D\text{RePr}/L)^{0.67}] \quad (4.32)$$

For turbulent flow in tubes without large viscosity variation (6),

$$hD/k = 0.023\text{Re}^{0.8}\text{Pr}^{0.33} \quad (4.33)$$

For turbulent flow in tubes with variable viscosity (4),

$$hD/k = 0.026\text{Re}^{0.8}\text{Pr}^{0.33}(\mu_b/\mu_w)^{0.14} \quad (4.34)$$

If flow is not fully developed, the recommended relation (7) is

$$hD/k = 0.036\text{Re}^{0.8}\text{Pr}^{0.33}(D/L)^{0.055} \quad (4.35)$$

when $10 < L/D < 400$.

For turbulent flow and noncircular cross sections, we can use eq 4.34 and 4.35 by substituting $4R_h$ (the hydraulic radius, defined by eq 3.37 to 3.39) for the tube diameter D.

Example 4.5 is a convective heat transfer problem. The solution involves finding an average temperature, determining the fluid properties (density, viscosity, etc.) at the temperature, calculating Re, Pr, and other dimensionless numbers, selecting the appropriate convection equation, and calculating the required information.

EXAMPLE 4.5. CALCULATING THE HEAT TRANSFER COEFFICIENT h

In a heat exchanger, water flows through a long copper tube (inside diameter 2.2 cm) with an average velocity of 2.13 m/s. The water is heated by steam condensing at 150 °C on the outside of the tube. Water enters at 15 °C and leaves at 60 °C. What is the heat transfer coefficient h for the water?

The average bulk temperature of the water is $(15 + 60)/2$ or 37.5 °C. The water properties at this temperature are ρ equals 993 kg/m³, C_p equals 4.17×10^3 J/kg-°C, μ equals 6.83×10^{-4} kg/m-s, and k equals 0.630 W/m-°C.

$$\text{Re} = (0.022 \text{ m})(2.13 \text{ m/s})(993 \text{ kg/m}^3)/6.83 \times 10^{-4} \text{ kg/m-s} = 68{,}100$$

$$\text{Pr} = (4.17 \times 10^3 \text{ J/k-°C})(6.83 \times 10^{-4} \text{ kg/m-s})/(0.630 \text{ W/m-°C}) = 4.53$$

Flow is turbulent, which means that

$$hD/k = 0.026\text{Re}^{0.8}\text{Pr}^{0.33}(\mu_b/\mu_w)^{0.14}$$

Because the steam is condensing at 150 °C and the average water bulk temperature is 37.5 °C, we can assume that the average temperature is 93.75 °C at the wall. Hence

$$\mu_w = 3.06 \times 10^{-4} \text{ kg/m-s}$$

Then

$$h = (0.026)(0.630)(68{,}100)^{0.8}(4.53)^{0.33}(6.83/3.06)^{0.14}/(0.022 \text{ m})$$

$$h = 10{,}087 \text{ W/m}^2\text{-°C}$$

In Example 4.5, the viscosity correction for water would not be large and could have been neglected.

Example 4.6 involves a significant viscosity change, which cannot be neglected. Example 4.6 also appears to be a case with too little information available. By use of thermodynamics, however, we can overcome the problem.

EXAMPLE 4.6. HEATING A VISCOUS FLUID

Oil enters a 1.25-cm-diameter, 3-m-long tube at 38 °C. The tube wall is maintained at 66 °C, and flow velocity is 0.3 m/s. What is the total heat transfer to the oil and the oil's exit temperature?

To solve this problem, we must use an assumed value for the exit temperature of the oil. The second law of thermodynamics shows us that the oil exit temperature cannot exceed 66 °C. The exit temperature must be between 38 °C and 66 °C. As a start, we will let this temperature be 50 °C. The average temperature of the fluid is $(50 + 38)/2$ or 44 °C. For this temperature,

$$\text{Re} = DV\rho/\mu = 18.2$$

and

$$\text{Pr} = C_p\mu/k = 2790$$

Flow is laminar, and the equation to find h is

$$hD/k = 1.86(\text{RePr})^{0.33}(D/L)^{0.33}(\mu_b/\mu_w)^{0.14}$$

$$h = (0.144/0.0125)(1.86)(18.2)^{0.33}(2790)^{0.33}(1/240)^{0.33}(671/20.3)^{0.14}$$

and

$$h = 199 \text{ W/m}^2\text{-}°C$$

From the first law of thermodynamics, assuming that there are no potential or kinetic energy changes and no shaft work,

$$Q = \Delta H$$

and

$$hA(T_w - T_{b, \text{avg}}) = (\dot{m}\, C_p)_{\text{oil}}(T_{\text{exit}} - T_{\text{in}})$$

Solving for the exit temperature,

$$T_{\text{exit}} = 46 °C$$

The difference between the value of h at the assumed exit temperature of 50 °C and 46 °C is small with negligible effect on the properties. Hence, h remains at 199 W/m²-°C. Solving for q,

$$q = 1178 \text{ W}$$

The next case illustrates a common calculation for heat exchanger design and analysis: estimating the effect of multiple thermal resistances to heat transfer and an overall heat transfer coefficient for a noncircular conduit.

EXAMPLE 4.7. CALCULATING *U* FOR AN ANNULAR HEAT EXCHANGER

In a heat exchanger, hot flue gases at 427 °C flow inside a 2.5-cm-diameter copper tube (0.16-cm-thick wall). A 5.0-cm tube is placed around the 2.5-cm tube and high-pressure water at 150 °C flows in the annular space. For a water flow rate of 1.5 kg/s and a total heat transfer of 17,600 W/h, what is the estimated heat exchanger length needed for a gas flow rate of 0.75 kg/s? We can assume that flue gas has properties similar to air.

$$\text{Re} = DV\rho/\mu$$

Multiplying Re by the area of a circle,

$$\text{Re} = (DV\rho/\mu)(\pi D^2/4)/(\pi D^2/4)$$

Because $V\rho(\pi D^2/4)$ equals the mass flow rate w,

$$\text{Re} = wD/(\pi D^2/4)\mu$$

Then,

$$\text{Re}_{\text{gas}} = (0.75 \text{ kg/s})(0.025 \text{ m})/(\pi/4)(0.025 \text{ m})^2(3.3 \times 10^{-5} \text{ kg/m-s})$$
$$\text{Re}_{\text{gas}} = 1{,}157{,}500$$

and

$$\text{Pr} = 0.684$$

To find h, we use the equation without viscosity correction because viscosity effects are small in a gas.

$$h_{\text{gas}} = (0.0524 \text{ W/m-}°\text{C})(0.023)(1.1575 \times 10^6)^{0.8}(0.684)^{0.33}/0.025 \text{ m}$$

and

$$h_{\text{gas}} = 3000 \text{ W/m}^2\text{-}°\text{C}$$

For the water, we must calculate the hydraulic diameter

$$D_{\text{hyd}} = 4(\pi D_2{}^2/4 - \pi D_1{}^2/4)/(\pi D_2 + \pi D_1) = D_2 - D_1$$
$$D_{\text{hyd}} = (0.05 - 0.0282) \text{ m} = 0.0218 \text{ m}$$

Then,

$$\text{Re}_{\text{water}} = (1.5 \text{ kg/s})(0.0218 \text{ m})/(0.00133 \text{ m}^2)(1.86 \times 10^{-4} \text{ kg/m-s})$$
$$\text{Re}_{\text{water}} = 132{,}000$$

and

$$\text{Pr}_{\text{water}} = 1.17$$

Then

$$h_{\text{water}} = (0.683 \text{ W/m-}°\text{C}/0.0218 \text{ m})(132{,}000)^{0.8}(1.17)^{0.33}(0.023)$$
$$h_{\text{water}} = 9480 \text{ W/m}^2\text{-}°\text{C}$$

The overall heat transfer is

$$q = UA\Delta T$$

and

$$L = (q/\Delta T\pi)[1/h_{\text{gas}}D_1 + (\ln D_o/D_i)/2k + 1/h_{\text{water}}D_o]$$
$$L = (17{,}600 \text{ W}/\pi277 \text{ °C})[1/(3{,}000 \text{ W/m}^2\text{-}°\text{C})(0.025 \text{ m}) +$$
$$(\ln 2.82/2.5)/2(368.8 \text{ W/m-}°\text{C}) + 1/(9{,}480 \text{ W/m}^2\text{-}°\text{C})(0.0282 \text{ m})]$$
$$L = 0.35 \text{ m}$$

There are three thermal resistances in this heat transfer problem. From best to worst conductors, the materials are the metal wall, the water, and the flue gas. A mix of resistances exists in most heat exchange systems. Because the gas is the worst conductor, we say that the gas side is the controlling factor in this heat transfer process.

HEAT TRANSFER WITH FLOW AROUND OBJECTS

Many cases of heat transfer occur with flow around objects. The heat transfer relation for flow over objects is modeled by flow around various geometrical cross sections.

For cylinders, heat transfer is given by the following relationship, with f subscripts indicating the properties determined at T_f, the average of the fluid's approach temperature and surface temperature of the cylinder (8, 9).

$$hD/k_f = C[VD\rho_f/\mu_f]^n Pr_f^{0.33} \tag{4.36}$$

Values for C and n are given in Table 4.3 for ranges of the Reynolds number Re_{d_f} for which the d_f subscript is a Reynolds number based on the cylinder diameter and the film temperature.

Equation 4.36 can also be applied to conduits of noncircular cross section. The C and n values are given in Table 4.4.

For gas flow around spheres and $17 < Re < 70,000$, the relationship is

$$hD/k_f = 0.37[VD\rho_f/\mu_f]^{0.6} \tag{4.37}$$

For liquid flow around spheres and $1 < Re < 200,000$,

$$hD/k_f = (1.2 + 0.53Re^{0.54})Pr^{0.3}(\mu/\mu_w)^{0.25} \tag{4.38}$$

where μ is the viscosity of the fluid.

When we have many tubes or cylinders in a system, we use a different method of finding C and n for eq 4.36. This method consists of using the geometry of the

TABLE 4.3 Constant Values for Equation 4.36, Heat Transfer Around Cylinders

Re_{d_f}[a]	C	n
0.4–4	0.989	0.330
4–40	0.911	0.385
40–4,000	0.683	0.466
4,000–40,000	0.193	0.618
40,000–400,000	0.0266	0.805

[a]The Reynolds number is based on the cylinder diameter and the film temperature.

TABLE 4.4   Constant Values for Equation 4.36, Heat Transfer
Around Noncircular Conduits

Geometry	Re	C	n
◇ D	5×10^3–10^5	0.246	0.588
□ D	5×10^3–10^5	0.102	0.675
⬡ D	5×10^3–1.93×10^4	0.160	0.638
	1.95×10^4–10^5	0.0385	0.782
⬡ D	5×10^3–10^5	0.153	0.638
│ D	4×10^3–1.5×10^4	0.228	0.731

tube bank, including the spacing between tube rows, S_p, the tube diameter, D, and
the vertical spacing between tubes, S_n (Figure 4.5). The tubes can be in-line or staggered.

In tube banks, the fluid's approach velocity is not the actual velocity in the bank
itself. The number of rows influences the velocity behavior. In practice, it turns out
that this influence occurs for up to 10 rows. Thereafter, the velocity effect remains

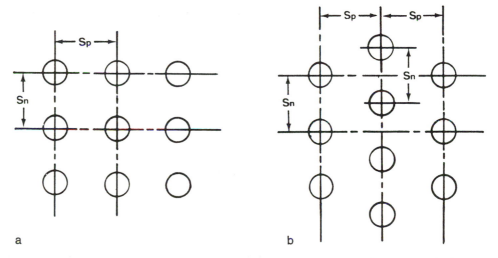

FIGURE 4.5.   Heat transfer in a tube bank is affected by the tube configuration, whether the
tubes are (a) in-line or (b) staggered. The values of S_p and S_n determine the constant C and the
exponential power n (see Table 4.5) to be used in eq 4.36. (Reproduced with permission from
reference 22. Copyright 1981 McGraw-Hill.)

the same. We correlate C and n for 10 or more rows with geometry (Table 4.5). If we have fewer than 10 rows, we use Table 4.6 to correct the value.

Pressure drops for flow of gases over tube banks are frequently needed. In these cases, we use the expression (13)

$$\Delta P = \frac{2f'G^2N}{\rho}\left(\frac{\mu_{\text{wall}}}{\mu_{\text{bulk}}}\right)^{0.14} \tag{4.39}$$

where G is the mass velocity (kg/m²-s) at minimum flow area and N is the number of transverse rows. The empirical factor f' for staggered tubes (13) is

$$f' = [0.25 + 0.118/\left|\left(\frac{S_n - D}{D}\right)\right|^{1.08}]\text{Re}^{-0.16} \tag{4.40}$$

and for in-line tubes is

$$f' = \left(0.044 + \frac{0.08\ S_p/D}{\left|\left(\frac{S_n - D}{D}\right)\right|^{0.43 + 1.13\ D/S_p}}\right)\text{Re}^{-0.15} \tag{4.41}$$

EXAMPLE 4.8. HEAT TRANSFER ACROSS IN-LINE TUBES

Energy is to be recovered by transferring heat from hot water flowing through tubes arranged in a bank to air flowing around the tubes. Six rows of tubes 15.24 m high are set up as an in-line arrangement. The tubes are 0.0064 m in diameter and $S_n = S_p = 0.0192$ m. The tube wall temperature is 93.3 °C. Atmospheric air is forced across the tube bank at an inlet velocity of 4.57 m/s. What is the total heat transferred per unit length of the tube bank?

$$S_p/D = 0.0192/0.004 = 3.0 = S_n/D$$

From Table 4.5, we find C equal to 0.317 and n equal to 0.608. Next we find T_f,

$$T_f = (93.3 + 21.1)/2 = 57.2\ °C$$

For this temperature and atmospheric pressure, the properties needed to calculate Re are

$$\mu_f = 1.97 \times 10^{-5}\ \text{kg/m-s}$$
$$k_f = 0.028\ \text{W/m-°C}$$
$$\rho_f = 1.0684\ \text{kg/m}^3$$
$$C_{pf} = 1.047\ \text{kJ/kg}$$

and

$$V_{\text{max}} = 4.57\ \text{m/s}\ [0.0192/(0.0192 - 0.0064)] = 6.86\ \text{m/s}$$
$$\text{Re} = (0.0064\ \text{m})(6.86\ \text{m/s})(1.0684\ \text{kg/m}^3)/1.97 \times 10^{-5}\ \text{kg/m-s}$$
$$\text{Re} = 2360$$

TABLE 4.5 Constants for Equation 4.36 for Banks of 10 or More Rows of Tubes in Heat Transfer Around Cylinders

S_n/D	S_p/D	1.25		1.5		2.0		3.0	
		C	n	C	n	C	n	C	n
In-line	1.25	0.386	0.592	0.305	0.608	0.111	0.704	.0703	0.752
	1.5	0.407	0.586	0.278	0.620	0.112	0.702	.0753	0.744
	2.0	0.464	0.570	0.332	0.602	0.254	0.632	0.220	0.648
	3.0	0.322	0.601	0.396	0.584	0.415	0.581	0.317	0.608
Staggered	0.6							0.236	0.636
	0.9					0.495	0.571	0.445	0.581
	1.0			0.552	0.558				
	1.125							0.575	0.560
	1.25	0.575	0.556	0.561	0.554	0.531	0.563	0.579	0.562
	1.5	0.501	0.568	0.511	0.562	0.576	0.556	0.542	0.568
	2.0	0.448	0.572	0.462	0.568	0.502	0.568	0.498	0.570
	3.0	0.344	0.592	0.395	0.580	0.535	0.556	0.467	0.574

NOTE: Blank sections indicate lack of data.

SOURCE: Reproduced with permission from reference 11. Copyright 1937 American Society of Mechanical Engineers.

TABLE 4.6   Ratio of h for N Rows Deep to h for 10 Rows Deep

Form	1	2	3	4	5	6	7	8	9	10
Staggered	0.68	0.75	0.83	0.89	0.92	0.95	0.97	0.98	0.99	1.0
In-line	0.64	0.80	0.87	0.90	0.92	0.94	0.96	0.98	0.99	1.0

SOURCE: Reproduced from reference 12. Copyright 1952.

Then, using eq 4.36,

$$h = (0.0284 \text{ W/m-°C}/0.0064 \text{ m})(0.317)(2360)^{0.608}(0.7)^{0.33}$$

$$h = 142 \text{ W/m}^2\text{-°C}$$

For six rows, the effective h (from Table 4.6) is

$$h = 0.94(142 \text{ W/m}^2\text{-°C}) = 133 \text{ W/m}^2\text{-°C}$$

Then,

$$q = hA\Delta T$$

$$q = (133 \text{ W/m}^2\text{-°C})(5.974 \text{ m}^2/\text{m})(72.2 \text{ °C})$$

and

$$q = 57{,}690 \text{ W/m}$$

HEAT TRANSFER OF A CONDENSING FILM

Frequently, heat transfer is accompanied by a phase change, such as boiling or condensation. In these instances, we use a heat transfer coefficient different from the usual h for convective cases. We first consider condensation.

If a cool vertical flat plate is exposed to a condensing vapor, a film forms and flows down the plate in laminar flow (Figure 4.6). A theoretical analysis of this situation yields a solution for the mean heat transfer coefficient h_m with laminar flow of the condensate (14),

$$h_m = 1.13 \, [k_f^3\rho_f^2 g\lambda/\mu_f L\Delta T]^{0.25} \tag{4.42}$$

or

$$h_m(\mu_f^2/k_f^3\rho_f^2 g)^{0.33} = 1.88(4\Gamma/\mu_f)^{-0.33} \tag{4.43}$$

In these equations, λ is the heat of vaporization, Γ is the mass flow rate per unit perimeter (the mass flow rate divided by width of a flat wall or the circumference of a tube), and ΔT is the temperature difference between the condensing vapor and the

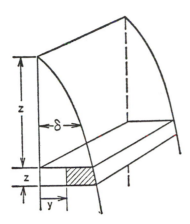

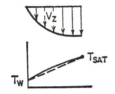

FIGURE 4.6. Condensing film. (Reproduced with permission from reference 23. Copyright 1961 Prentice-Hall.)

surface. The subscript f indicates that the property is to be evaluated at the film temperature. In this case the film temperature T_f is not the average of T_{wall} and $T_{\text{condensation}}$, but rather $T_{\text{condensation}} - 3/4(T_{\text{condensation}} - T_{\text{wall}})$. Equations 4.42 and 4.43 are both for laminar condensate flow (i.e., $4\Gamma/\mu_f < 1800$).

When $4\Gamma/\mu_f$ exceeds 1800, the heat transfer coefficient can be obtained (*15, 16*) from

$$h_m(\mu_f^2/k_f^3\rho_f^2 g)^{0.33} = 0.0077(4\Gamma/\mu_f)^{0.4} \tag{4.44}$$

For condensation or vaporization on horizontal tubes, when there is a vertical tier of N such horizontal tubes, the heat transfer coefficient is (*17*)

$$h_m = 0.725(k_f^3\rho_f^2 g\lambda/ND_{\text{tube}}\mu_f\Delta T)^{0.25} \tag{4.45}$$

or

$$h_m = 0.95(k_f^3\rho_f^2 gL/\mu_f W)^{0.33} \tag{4.46}$$

where W is mass flow rate of condensate.

Usually, the condensate flow for horizontal tubes is laminar. If, however, $2\Gamma'/\mu_f$ exceeds 2100 (where Γ' is W/L), then eq 4.44 should be modified to determine h_m by

$$h_m(\mu_f^2/k_f^3\rho_f^2 g)^{0.33} = 0.0077(2\Gamma'/\mu_f)^{0.4} \tag{4.47}$$

HEAT TRANSFER OF CONDENSING DROPLETS

If we suppose that a condensing vapor forms droplets on a surface, considerably higher heat transfer coefficients (four to eight times higher than for a condensing film) are obtained. Heat transfer equipment can be coated to promote droplet condensation and, thereby, enhance heat transfer. Because this condition, known as drop-wise condensation, cannot be maintained continuously, film condensation is used as a design basis.

HEAT TRANSFER DUE TO VAPORIZATION

Another type of phase change occurs with heat transfer when a liquid boils. Boiling is a complex phenomenon (Figure 4.7). In the figure, the energy flux q/A is plotted against the difference in temperature between the wall and the fluid's saturation temperature. The ordinate is equivalent to plotting the heat transfer coefficient against ΔT.

We can qualitatively replicate the behavior shown in Figure 4.7 by performing a simple experiment consisting of an immersion heater placed into a beaker filled with water. When the heater is turned on, we first observe motion in the water (free convection). Next bubbles rise from the heating surface, slowly at first, then more rapidly (this is called nucleate boiling). At some point, the bubbles at the heating surface begin to coalesce (unstable film) and ultimately form a vapor film at the heating surface (stable film). Finally, the heat transfer coefficient increases because of radiative heat transfer.

The peaks of the heat flux and the stable film in Figure 4.7 correspond to drop-wise and film condensation, respectively. The heat transfer regimes most useful for process work are II and III. Regime VI is not commonly encountered or desired.

Because each step of boiling has a distinct character, a different heat transfer equation must be used for each type of boiling. Each equation is fixed by the difference between the surface T_w and saturation T_{sat} temperatures.

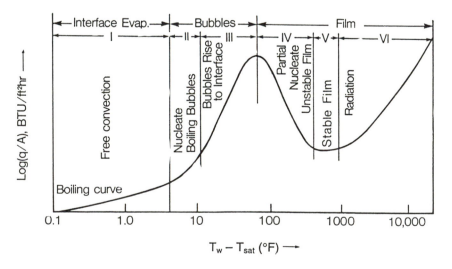

FIGURE 4.7.   As boiling progresses, the liquid at the heat transfer surface goes through several stages, as shown by the regimes I through VI. In the first stage we see convective movement of the liquid. In stages II and III, bubbles form. In stage IV, a film develops at the surface, and in stage V, the film is fully developed and stable. In industrial processes, stages II and III are the most useful. (Reproduced with permission from reference 23. Copyright 1961 Prentice-Hall.)

The first of these equations covers free convection where $(T_w - T_{sat}) < 5.0$ °F. This regime is one of interfacial evaporation (*18*), where

$$\text{Nu} = 0.61\text{Pr}^{0.25}\text{Gr}^{0.25} \tag{4.48}$$

In the regime of nucleate boiling where bubbles form and rise to the fluid surface, heat transfer data can be correlated to the Reynolds and Prandtl numbers as in single-phase forced convection.

$$\text{Nu} = \phi\ (\text{Re, Pr}) \tag{4.49}$$

A typical correlation for nucleate boiling (*19*) is

$$C_{p_l}(T_w - T_{sat})/\lambda = C'[(q/A)/\mu_l\lambda\ \{g_c\sigma/g(\rho_l - \rho_g)\}^{0.5}]^{0.33}\ \text{Pr}^{1.7} \tag{4.50}$$

The l and g subscripts refer to liquid and gaseous phases, σ is the surface tension, and C' is an empirical constant dependent on the type of surface. A list of C' values for various fluid–surface combinations is given in Table 4.7.

If there is convection as well as nucleate boiling, then the total flux is

$$(q/A)_{total} = (q/A)_{moderate\ boiling} + (q/A)_{convection} \tag{4.51}$$

The convection flux is determined by the appropriate convective equation.

The peak heat flux (the maximum of Figure 4.7) can be estimated (*20*) with the relation

$$(q/A)_{max}/\rho_g\lambda = 143g^{0.25}[(\rho_l - \rho_g)/\rho_g]^{0.6} \tag{4.52}$$

where g is the acceleration of gravity in Gs (a multiple of 32.2 ft/s^2).

TABLE 4.7 Nucleate Boiling Values of C' for Equation 4.50 for Various Liquid-Heating Surface Combinations

Combination	C'
Water and copper	0.013
Water and platinum	0.013
Water and brass	0.0060
n-Butyl alcohol and copper	0.00305
Isopropyl alcohol and copper	0.00225
n-Pentane and chromium	0.015
Benzene and chromium	0.010
Ethyl alcohol and chromium	0.027

Finally, for a stable film boiling on a tube, the heat transfer coefficient is a combination of a conduction h and a radiative h,

$$h = h_{\text{conduction}} + h_{\text{radiative}} \qquad (4.53)$$

$$h_{\text{conduction}} = 0.62[k_g^3 \rho_g(\rho_l - \rho_g)g(\lambda + 0.4C_g\Delta T)/D_{\text{tube}}\mu_g\Delta T]^{0.25} \qquad (4.54)$$

and

$$h_{\text{radiative}} = \sigma\epsilon(T_w^4 - T_{\text{sat}}^4)/T_w - T_{\text{sat}}$$

where σ is the Stefan–Boltzmann constant ($\sigma = 0.1714 \times 10^{-8}$ Btu/h-ft²-°R⁴) and ϵ is the emissivity of the surface (*21*).

The two following examples consider condensation on vertical and horizontal systems. In the vertical case we have turbulent flow, which is not unusual for a vertical surface. The magnitude of the flow rate in the second case is not immediately apparent, but turbulent flow is rare in cases of condensation for horizontal arrangements, so that we assume laminar flow to solve for h_m.

EXAMPLE 4.9. CONDENSATION HEAT TRANSFER

What length of 0.0508-m outside-diameter vertical tube is required to condense 0.1827 kg/s of saturated steam at 127.8 °C if the tube wall temperature is 72.2 °C?

$$\Gamma = W/\pi D = 0.1827 \text{ kg/s}/\pi 0.0508 \text{ m} = 1.141 \text{ kg/s-m}$$

$$T_f = 127.8 \text{ °C} - (3/4)(127.8 - 72.2) \text{ °C} = 85.1 \text{ °C}$$

$$\mu_f = 0.00033 \text{ kg/m-s}$$

Then

$$\text{Re} = 4\Gamma/\mu_f = 4(1.141 \text{ kg/s-m})/0.00033 \text{ kg/m-s} = 13{,}800$$

Thus, condensate flow is turbulent and

$$h_m = 0.0077\text{Re}^{0.4}(k_f^3\rho_f^2 g/\mu_f^2)^{0.33}$$

$$h_m = 0.0077(13{,}800)^{0.4}[(0.673 \text{ W/m-°C})^3(869.78 \text{ kg/m}^3)^2(9.8 \text{ m/s}^2)/$$
$$(0.00033 \text{ kg/m-s})^2]^{0.33}$$

$$h_m = 9581 \text{ W/m}^2\text{-°C}$$

and

$$q = W\lambda = (0.183 \text{ kg/s})(2179.3 \text{ kJ/kg})$$

$$q = 398{,}575 \text{ W}$$

$$A = (398{,}575 \text{ W})/(9581 \text{ W/m}^2\text{-°C})(55.6 \text{ °C}) = 0.748 \text{ m}^2$$

and

$$L = 0.748 \text{ m}^2/\pi 0.051 \text{ m} = 4.67 \text{ m}$$

EXAMPLE 4.10. CONDENSATION ON A BANK OF TUBES

What is the value of h_m for dry saturated steam at 100 °C condensing outside a bank of horizontal tubes, 16 tubes high? The average temperature of the outer tube surface is 93 °C, and the tube outer diameter is 2.5 cm.

$$T_f = 100 - (3/4)(7) = 94.75 \ °C$$

Assuming laminar flow,

$$h_m = 0.725(k_f^3 \rho_f^2 g\lambda / ND_{\text{o.d. tube}}\mu_f^2\Delta T)^{0.25}$$

$$h_m = 0.725[(0.680 \ \text{W/m-°C})^3(859 \ \text{kg/m}^3)^2(9.81 \ \text{m/s}^2)(2.27 \times 10^6 \ \text{J/kg})/$$
$$16(0.025 \ \text{m})(0.0003 \ \text{kg/m-s})(7 \ °\text{C})]^{0.25}$$

$$h_m = 6419 \ \text{W/m}^2\text{-°C}$$

Checking the assumption of laminar condensate flow,

$$A = 16\pi D_{\text{o.d. tube}}L = 16\pi(0.025 \ \text{m})L$$
$$A = 1.26L \ \text{m}^2$$

and

$$w = h_m A\Delta T/\lambda = (6419 \ \text{W/m}^2\text{-°C})(1.26L \ \text{m}^2)(7 \ °\text{C})/2.27 \times 10^6 \ \text{J/kg}$$
$$w = 0.025L \ \text{kg/s}$$
$$\Gamma' = w/L = 0.025L \ \text{kg/s}/L \ \text{m} = 0.025 \ \text{kg/m-s}$$

so that

$$2\Gamma'/\mu_f = (2)(0.025 \ \text{kg/m-s})/0.0003 \ \text{kg/m-s} = 167$$

This flow value is less than 2100 and hence flow is laminar.

RADIATIVE HEAT TRANSFER

There are many types of electromagnetic radiation, one of which is thermal radiation. If radiation is conceived of as a "photon gas," then, from thermodynamics the energy density of the radiation is

$$E_b = \sigma T^4 \tag{4.55}$$

where σ is the Stefan–Boltzmann constant (5.66×10^{-8} W/m²-K⁴) and E_b is the radiation from a black body.

A black body is one in which both the absorptivity α and the emissivity ϵ are unity. For any body

$$\epsilon = \alpha \tag{4.56}$$

HOW AN OBJECT'S SHAPE AFFECTS RADIATION

When calculating radiative heat transfer, the system geometry—the surface configuration of the radiating body and the absorbing body is important. We must calculate how much the opposing surfaces "see" each other to determine the amount of radiative heat transfer between them. If this relationship is not known, we get erroneous results for radiative heat transfer. The area of transfer, A, is modified by a radiation shape factor F, also called a view factor, angle factor, or configuration factor, and is defined as

F_{12} = fraction of energy leaving surface 1 and reaching surface 2,
F_{21} = fraction of energy leaving surface 2 and reaching surface 1, and
F_{mn} = fraction of energy leaving surface m and reaching surface n.

For radiation between two smooth geometric surfaces (Figure 4.8), the radiation shape factors are derived in general from

$$A_1 F_{12} = A_2 F_{21} = \int_{A_1} \int_{A_2} \cos \theta_1 \cos \theta_2 \, dA_1 dA_2 / \pi r_{12}^2 \qquad (4.57)$$

where θ_1 and θ_2 are the angles respectively, between the ray r_{12} and the normals to dA_1 and dA_2. Figures 4.9 and 4.10 give radiation shape factors as functions of system geometry for rectangles and directly opposed planes. These plots can be used to determine the appropriate factor needed for radiative heat transfer by

$$q_{\text{net}} = A_1 F_{12}(E_{b1} - E_{b2}) = A_2 F_{21}(E_{b1} - E_{b2}) \qquad (4.58)$$

If the bodies undergoing radiant heat transfer are not black bodies and are connected by a third surface that does not exchange heat, then

$$q_{\text{net}} = \frac{\sigma A_1 (T_1^4 - T_2^4)}{\left(\dfrac{A_1 + A_2 - 2A_2 F_{12}}{A_2 - A_1 F_{12}^2} \right) + \left(\dfrac{1}{\epsilon_1} - 1 \right) + \dfrac{A_1}{A_2} \left(\dfrac{1}{\epsilon_2} - 1 \right)} \qquad (4.59)$$

where ϵ_1 and ϵ_2 are the emissivities of surfaces 1 and 2, respectively.

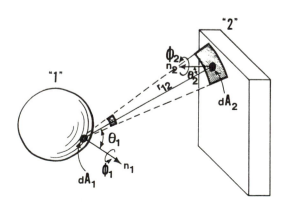

FIGURE 4.8. The quantity of heat transferred by radiation depends on a system's geometry. (Reproduced with permission from reference 24. Copyright 1960 Wiley.)

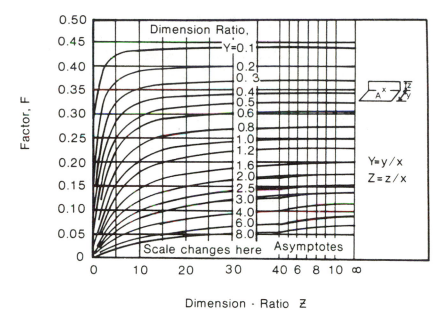

FIGURE 4.9. Radiation shape factor charts, such as this one for adjacent rectangles, are found from eq 4.58. (Reproduced with permission from reference 14. Copyright 1954 McGraw-Hill.)

For the special case of radiation between infinite parallel planes, the relationship becomes

$$q_{net} = \frac{\sigma A_1 (T_1^4 - T_2^4)}{\frac{1}{\epsilon_1} + \left(\frac{A_1}{A_2}\right)\left(\frac{1}{\epsilon_2} - 1\right)} \tag{4.60}$$

It is apparent from the derivation just seen that calculations for radiative heat transfer can become complex. Certain gases (CO_2, H_2O, SO_2) present between the objects can also influence radiation heat transfer. The following examples will give us some insight into radiative heat transfer.

EXAMPLE 4.11. RADIATION BETWEEN PARALLEL BLACK BODIES

Two parallel black plates, each of area 1.5×3 m, are spaced 1.5 m apart. One plate is maintained at 538 °C and the other at 260 °C. What is the net radiant energy interchange?

We can find F by using Figure 4.10, for which the ratios are

$$Y/D = 1.5/1.5 = 1.0$$
$$X/D = 3/1.5 = 2.0$$

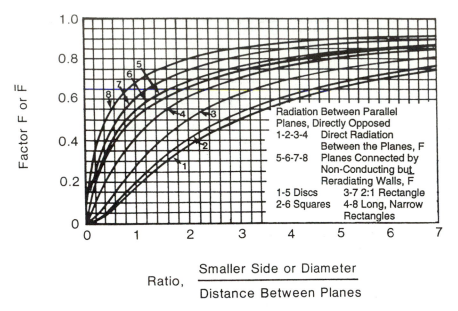

FIGURE 4.10. The shape of a radiative heat transfer system determines the shape factor F from eq 4.58 and its derivations. This chart is for parallel planes, discs, squares, and rectangles. (Reproduced with permission from reference 14. Copyright 1954 McGraw-Hill.)

We find the appropriate view factor from Figure 4.10. This case is direct radiation between planes. We use curves 1 through 4. The geometry is a 2-to-1 rectangle, which means that we use curve 3 with a ratio of 1.0 (1.5 m/1.5 m). Then

$$F_{12} = 0.285$$

$$q_{net} = A_1 F_{12}(E_{b1} - E_{b2}) = A_1 F_{12}\sigma(T_2^4 - T_1^4)$$

$$q_{net} = (4.5 \text{ m}^2)(0.285)(5.66 \times 10^{-8} \text{ W/m}^2\text{-K}^4)[(811 \text{ K})^4 - (533 \text{ K})^4]$$

and

$$q_{net} = 25,544 \text{ W}$$

EXAMPLE 4.12. RADIATION TO A CYLINDER

An oxidized steel tube ($\epsilon = 0.6$, outside diameter = 8 cm) passes through a silica brick furnace ($\epsilon = 0.8$, inside dimensions 15 cm × 15 cm × 15 cm). The temperature of the inside wall of the furnace is 1000 °C, and the outside of the tube is 555 °C. What is the rate of heat transfer?

The equation to use is for surfaces enclosed by a nonconducting but reradiating surface

$$q_{net} = \frac{\sigma A_1(T_1^4 - T_2^4)}{\left(\dfrac{A_1 + A_2 - 2A_2 F_{12}}{A_2 - A_1 F_{12}^2}\right) + \left(\dfrac{1}{\epsilon_1} - 1\right) + \dfrac{A_1}{A_2}\left(\dfrac{1}{\epsilon_2} - 1\right)}$$

In this case every bit of tube "sees" the furnace. Hence,

$$(A_1 + A_2 - 2A_2F_{12})/(A_2 - A_1F_{12}{}^2) = 1/1$$
$$A_1 = (\pi)(0.08 \text{ m})(0.15 \text{ m}) = 0.038 \text{ m}^2$$

and

$$A_2 = 6(0.15 \text{ m})^2 - 2(\pi/4)(0.08 \text{ m})^2 = 0.125 \text{ m}^2$$

Then,

$$q_{net} = (5.66 \times 10^{-8} \text{ W/m}^2\text{-K}^4)(0.038 \text{ m}^2)[(1273 \text{ K})^4 - (828 \text{ K})^4]/$$
$$[1/1 + (1/0.6 - 1) + (0.038 \text{ m}^2/0.125 \text{ m}^2)(1/0.8 - 1)]$$
$$q_{net} = 2649 \text{ W}$$

AN INTRODUCTION TO HEAT EXCHANGERS

The heat exchanger is a process device for interchanging heat between two fluids. Heat exchangers can have various geometries and configurations (Figures 4.11 to 4.16).

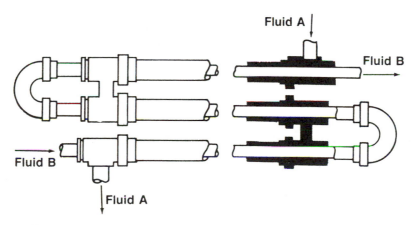

FIGURE 4.11.   Flow is countercurrent in this double-pipe heat exchanger.

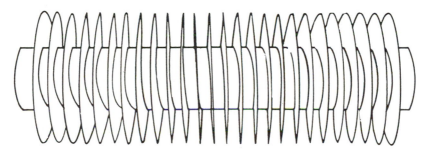

FIGURE 4.12.   A finned tube enhances heat exchange by increasing the area available for conduction.

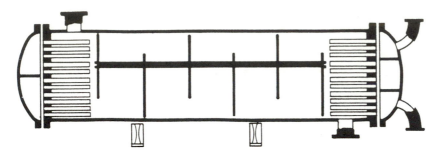

FIGURE 4.13.   A shell-and-tube heat exchanger has baffles on the shell side to enhance mixing of the shell fluid and can be operated either cocurrently or countercurrently. Both heads of this exchanger, which are filled with the tube liquid during operation, can be removed for inspection and cleaning. (Reproduced with permission from reference 18. Copyright 1964 Elsevier.)

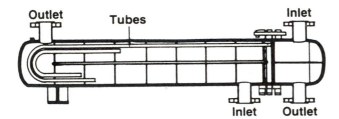

FIGURE 4.14.   A shell-and-tube exchanger with bends of tubing at one end has more area available for heat exchange than the one shown in Figure 4.13, but it is harder to clean. (Reproduced with permission from reference 18. Copyright 1964 Elsevier.)

CRITERIA FOR DESIGNING HEAT EXCHANGERS

The two most important criteria for designing or specifying heat exchangers are the amount of heat transfer surface required and the ratio of the inlet and outlet temperatures. Additional factors to be considered are the materials of construction, pressure drops across the exchanger due to friction, deposits and fouling, ease of maintenance, and economic feasibility of the exchanger purchase.

Heat exchangers can be attacked by corrosive fluids. Solutions to the corrosion problem include the use of alloys, coated surfaces, or nonmetallics. The physical configuration is also important because it can affect the flow pattern and cause frictional losses during operation. If a heat exchanger does the job thermally, but creates high pressure drops, the design is unacceptable.

Another criterion for heat exchanger design is ease of maintenance. A complicated unit that makes maintenance difficult can cause process problems. All of the design factors must be combined to optimize capital and operational costs.

In any heat-exchange system, deposits can form on solid surfaces. These deposits will add another resistance to the system and influence the overall heat-transfer coefficient. The change in the coefficient is expressed as a fouling factor

$$R_{\text{fouling}} = 1/U_{\text{fouled}} - 1/U_{\text{clean}} \qquad (4.61)$$

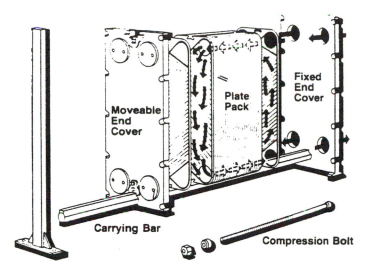

FIGURE 4.15. Each plate in a plate exchanger has a raised or rolled surface that creates high Reynolds numbers at low volume flow rates. This surface makes a plate exchanger particularly good at handling viscous or solid-laden fluids. A plate exchanger can also handle heat exchange between more than two fluids by routing fluids selectively through different sections. (Reproduced with permission from reference 18. Copyright 1964 Elsevier.)

Typical fouling factors (Table 4.8) can be determined by plotting the behavior of $1/U$ vs. flow rate periodically for a length of time after the unfouled exchanger is placed in service.

DESIGNING HEAT EXCHANGERS—COMPENSATING FOR TEMPERATURE AND CONFIGURATIONS

So far we have examined the nature of energy exchange from three modes of heat transfer and calculated an overall heat-transfer coefficient for steady-state heat transfer. We must also consider the effects on heat transfer of the flow pattern through

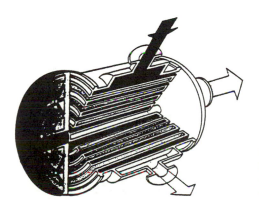

FIGURE 4.16. This spiral heat exchanger is completely made of glass. It is used to handle highly corrosive fluids.

TABLE 4.8 Fouling Factors for Heat Exchangers

Fluid	Fouling factor ($m^2\cdot°C/W$)
Seawater < 51.7 °C	9×10^{-5}
Seawater > 51.7 °C	2×10^{-4}
Oil	$7–9 \times 10^{-4}$
Refrigerant	2×10^{-4}

the exchanger and its configuration. The fluids interchanging energy in a double-pipe heat exchanger (pipe within a pipe) can flow either cocurrently (parallel) or countercurrently. Each flow pattern creates a distinct temperature profile over the length of the exchanger (Figure 4.17).

To account for the flow pattern, we use a differential approach to calculate the heat transfer. For a double-pipe heat exchanger

$$dq = U(T_b - T_c)dA \tag{4.62}$$

and

$$dq = (-WC_p)_{\text{hot}}dT_b = (WC_p)_{\text{cold}}dT_c \tag{4.63}$$

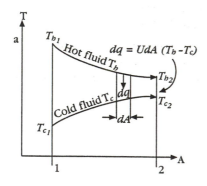

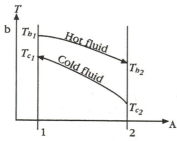

FIGURE 4.17. Cocurrent (a) and countercurrent (b) flows through a double-pipe heat exchanger have different temperature profiles. (Reproduced with permission from reference 22. Copyright 1981 McGraw-Hill.)

where W is mass flow rate, and the h and c subscripts indicate the initial temperature of the hot and cold flow. If we equate the dq in eq 4.62 and 4.63 and solve,

$$d(T_h - T_c)/(T_h - T_c) = -U[1/(WC_p)_{\text{hot}} + 1/(WC_p)_{\text{cold}}]dA \qquad (4.64)$$

Then,

$$\ln\left(\frac{T_{h_2} - T_{c_2}}{T_{h_1} - T_{c_1}}\right) = -UA\left(\frac{1}{(WC_p)_{\text{hot}}} + \frac{1}{(WC_p)_{\text{cold}}}\right) \qquad (4.65)$$

and finally,

$$(WC_p)_{\text{hot}} = \frac{q}{T_{h_1} - T_{h_2}} \qquad (4.66)$$

$$(WC_p)_{\text{cold}} = \frac{q}{T_{c_2} - T_{c_1}} \qquad (4.67)$$

and

$$q = UA\left\{\frac{(T_{h_2} - T_{c_2}) - (T_{h_1} - T_{c_1})}{\ln\left(\dfrac{T_{h_2} - T_{c_2}}{T_{h_1} - T_{c_1}}\right)}\right\} \qquad (4.68)$$

The temperature term in the brackets is known as the log mean temperature difference, ΔT_{lm}. For a double-pipe heat exchanger,

$$q = UA\Delta T_{\text{lm}} \qquad (4.69)$$

Equation 4.69 is applicable to any exchanger configuration, for example, cross flow or shell and tube, by using a correction factor F,

$$q = UAF\Delta T_{\text{lm}} \qquad (4.70)$$

F has been plotted for a shell-and-tube exchanger with one shell and any even number of tube phases (four, eight, etc.) (Figure 4.18).

If the log mean temperature difference for a desired exchanger is known, the exchanger can be sized using a chart (such as Figure 4.18) combined with eq 4.70. If the ΔT_{lm} is not known, sizing can become a process of trial and error. To avoid lengthy calculations, we can use a technique based on effectiveness defined as

$$\epsilon = \text{effectiveness} = \text{actual heat transfer/maximum possible heat transfer} \qquad (4.71)$$

For parallel flow heat exchangers,

$$q = (WC_p)_h(T_{h_1} - T_{h_2}) = (WC_p)_c(T_{c_2} - T_{c_1}) \qquad (4.72)$$

and for countercurrent flow

$$q = (WC_p)_h(T_{h_1} - T_{h_2}) = (WC_p)_c(T_{c_1} - T_{c_2}). \qquad (4.73)$$

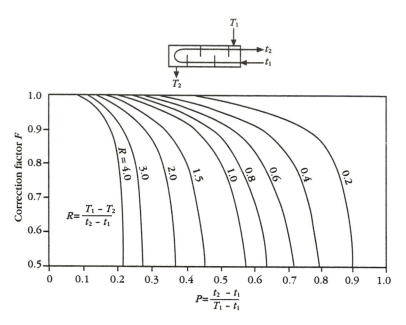

FIGURE 4.18. When using an exchanger with various configurations, the log mean temperature heat transfer (eq 4.68) needs to be corrected with a factor F. Values for the factor come from charts, such as this one, for a shell-and-tube heat exchanger with one shell pass and any even multiples of tube passes. (Reproduced with permission from reference 22. Copyright 1981 McGraw-Hill.)

We can define one fluid as having a maximum temperature change. This fluid will then have a minimum value of WC_p because of the energy balance.

Then,

$$q_{\text{maximum}} = (WC_p)_{\text{minimum}}(T_{h_{\text{inlet}}} - T_{h_{\text{outlet}}}) \tag{4.74}$$

Based on equation 4.74, the effectiveness ϵ is (for parallel flow)

$$\epsilon_h = (T_{h_1} - T_{h_2})/(T_{h_1} - T_{c_1}) \tag{4.75}$$

$$\epsilon_c = (T_{c_2} - T_{c_1})/(T_{h_1} - T_{c_1}) \tag{4.76}$$

Manipulation of eq 4.65 together with the effectiveness approach gives us the following type of solution

$$\epsilon = \{1 - \exp[-UA/(1/(WC_p)_{\text{min}} + 1/(WC_p)_{\text{max}})]\}/[1 + (WC_p)_{\text{min}}/(WC_p)_{\text{max}}] \tag{4.77}$$

The overall result is a series of solutions for ϵ in terms of the ratio of $(WC_p)_{\text{min}}/(WC_p)_{\text{max}}$ and $AU/(WC_p)_{\text{min}}$ for parallel, counterflow, and cross-flow systems (Figures 4.19–4.21).

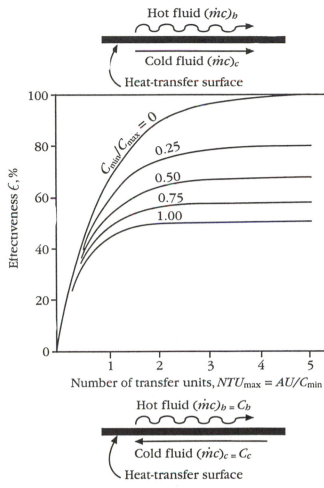

FIGURE 4.19. When we do not know the log mean temperature difference needed to solve eq 4.68, we can use a technique based on the supposed effectiveness ϵ of a particular heat exchanger configuration. The values shown here are for a cocurrent flow exchanger for different values of $(WC_p)_{min}$ or C_{min} and $(WC_p)_{max}$ or C_{max} (see eq 4.80). (Reproduced with permission from reference 22. Copyright 1981 McGraw-Hill.)

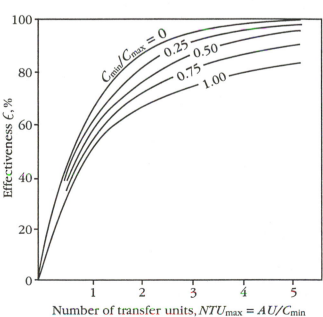

FIGURE 4.20. Effectiveness values for a countercurrent flow heat exchanger. (Reproduced with permission from reference 22. Copyright 1981 McGraw-Hill.)

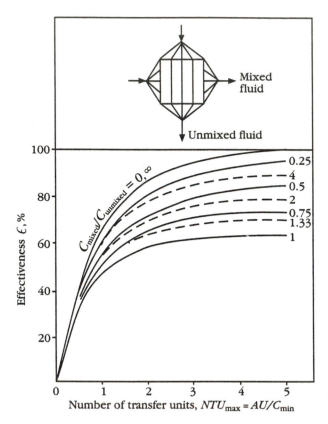

FIGURE 4.21. Effectiveness values for a cross-flow heat exchanger with one fluid mixed. (Reproduced with permission from reference 22. Copyright 1981 McGraw-Hill.)

REFERENCES

1. Carslaw, H. S.; Jaeger, J. C. *Conduction of Heat in Solids;* Oxford University: London, England, 1959.
2. Crank, J. *The Mathematics of Diffusion;* Oxford University: London, England, 1959.
3. Gurney, H. P.; Lurie, J. *Ind. Eng. Chem.* **1923,** *15,* 1170.
4. Sieder, E. N.; Tate, C. E. *Ind. Eng. Chem.* **1936,** *28,* 1429.
5. Hausen, H. V. *D. I. Z.* **1943,** *4,* 91.
6. Colburn, A. P. *Purdue U. Eng. Bull.* **1942,** *26,* 1.
7. Nusselt, W. *Forsch. Geb. Ingenieurwes.* **1931,** *2,* 309.
8. Hilpert, R. *Forsch. Geb. Ingenieurwes.* **1933,** *4,* 220.
9. Knudsen, J. D.; Katz, D. L. *Fluid Dynamics and Heat Transfer;* McGraw-Hill: New York, NY, 1958.
10. Jakob, M. *Heat Transfer;* John Wiley and Sons: New York, 1949.
11. Grimson, E. D. *Trans. ASME* **1937,** *59,* 583.
12. Kays, W. M.; Lo, R. K. *Stanford U. Tech. Rep. 15;* Navy Contract N6-ONR251 T.O.6, 1952.
13. Jakob, M. *Trans. ASME* **1938,** *60,* 384.
14. McAdams, W. H. *Heat Transmission*, 3rd ed.; McGraw-Hill: New York, 1954.
15. Badger, W. L. *Trans. Am. Inst. Chem. Eng.* **1937,** *33,* 441.
16. Kirkbride, C. G. *Ind. Eng. Chem.* **1934,** *26,* 425.
17. Nusselt, W. *V. D. I. Z.* **1916,** *60,* 541.
18. Coulson, J. M.; Richardson, J. F. *Chemical Engineering;* Pergamon: New York, 1964; Vol. 1.

19. Rohsenow, W. M. *Trans. ASME* **1952,** *74,* 969.
20. Rohsenow, W. M.; Griffith, P. *Joint National Heat Transfer Conference;* Louisville, KY, March 1955.
21. Bromley, L. A. *Chem. Eng. Prog.* **1950,** *46,* 221.
22. Holman, J. P. *Heat Transfer,* 5th ed.; McGraw-Hill: New York, 1981, out of print.
23. Rohsenow, W. M.; Choi, H. *Heat, Mass and Momentum Transfer;* Prentice-Hall: Englewood Cliffs, NJ, 1961.
24. Bird, R. B.; Stewart, W. E.; Lightfoot, E. N. *Transport Phenomena;* John Wiley and Sons: New York, 1960.

FURTHER READING

Bennett, C. O; Myers, J. E. *Momentum, Heat and Mass Transfer,* 3rd ed.; McGraw-Hill: New York, 1982.
Coulson, J. M.; Richardson, J. F. *Chemical Engineering;* Pergamon: London, England, 1978; Vol. 1 and 2.
Fahien, R. *Fundamentals of Transport Phenomena;* McGraw-Hill: New York, 1983.
Foust, A. S.; Wenzel, L. A.; Clump, C. W.; Maus, L.; Anderson, L. *Principles of Unit Operations,* 2nd ed.; reprint of 1980 ed.; Krieger, 1990.
Geankoplis, C. J. *Transport Processes and Unit Operations;* Allyn and Bacon: Boston, MA, 1978.
Greenkorn, R. A.; Kessler, D. P. *Transfer Operations;* McGraw-Hill: New York, 1972, out of print.
Holman, J. P. *Heat Transfer,* 7th ed.; McGraw-Hill: New York, 1989.
Jakob, M. *Heat Transfer;* John Wiley and Sons: New York, 1949, out of print.
Knudsen, J. D.; Katz, D. L. *Fluid Dynamics and Heat Transfer;* McGraw-Hill: New York, 1958, out of print.
McAdams, W. H. *Heat Transmission,* 3rd ed.; reprint of 1954 ed., Krieger: New York, 1985.
McCabe, W. L.; Smith, J. C.; Harriott, P. *Unit Operations of Chemical Engineering;* McGraw-Hill: New York, 1985.

5

Mass Transfer

Heterogeneous systems, purification and separation needs, processes, and environmental constraints all combine to make mastery of mass transfer a requirement for the proper handling of chemical and petroleum processes. We must understand the mechanisms of mass transfer in distillations, filtrations, centrifugations, absorptions, and many other separation and purification operations that the chemical industry uses.

Mass transfer is the movement of a chemical species from a region of higher to lower concentration. We can see the effect of mass transfer when colored solid crystals dissolve in a clear liquid, air pollutants disperse, and materials dry. Mass transfer is achieved by concentration differences that give rise to concentration gradients (changes of concentration over a distance). In Chapter 1 we showed that the mass or molar flux was related to the concentration gradient by the system's diffusivity.

Concentration gradients are not the only driving forces for mass transfer. Mass transfer can also be driven by differences in pressure (as in centrifugations), external forces (electrolytic cells, Figure 5.1) or temperature (thermal diffusion). To understand mass transfer, therefore, we must examine the interplay between the different driving forces and fluxes.

As described in Chapter 1, mass transfer, as governed by Fick's law of diffusion, is a vector (diffusion has three components). The same is true of heat transfer, which also has three components. However, fluid flow, which involves momentum transfer, is quite different because it has nine components and is known as a tensor.

To develop an understanding of the interplay of the different fluxes and the gradients associated with them, we can refer to the work of Nobel Laureate Lars Onsager (*1,2*). Onsager assigned each flux a degree commensurate with its mathematical complexity. Fluid flow, as a tensor, was assigned second degree. Both heat transfer and mass transfer, as vectors, were assigned the first degree. Finally, chemical reaction, a scalar entity, was assigned the zeroth degree. The fluxes were then compared (Table 5.1) to their respective gradients; chemical reaction is brought about by chemical affinity. Onsager stated that fluxes of the same degree were analogous and could be driven by the alternate gradient. Heat and mass transfer, for example, are analo-

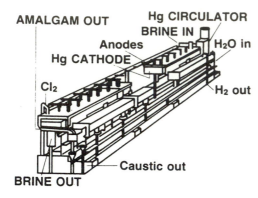

FIGURE 5.1. The operation of a mercury cell is a good example of mass transfer driven by external forces rather than concentration gradients.

gous because they are vectors, so mass transfer can be achieved by a temperature gradient (called thermal diffusion or the Soret effect). The alternate process, heat transfer by mass transfer gradient, occurs because of energy developed in the mass transfer process. Onsager termed the relationship between like-degree fluxes "coupling."

Onsager also postulated that a second-degree flux, such as momentum transfer, and a zeroth degree entity, such as chemical reaction (fluxes differing by two degrees), could also couple, although his second type of coupling has not been demonstrated experimentally. All in all then, mass transfer can be driven by several gradients, each of which we shall examine.

TABLE 5.1.   Flux–Gradient Interactions

Flux	Velocity Gradient	Temperature Gradient	Concentration Gradient	Chemical Affinity
Momentum (tensor, 9 components)	$\tau_{yx} = \mu\, dv/dy$			
Heat (vector, 3 components)		$q_y = k\, dT/dy$ Fourier's law	Dufour effect	
Mass (vector, 3 components)		Soret effect, thermal diffusion	$J_{A_y} = D_{AB}\, dC_A/dy$ Fick's law	
Chemical reaction (scalar)				rate $= k_r C_A^n$

NOTE: All equations shown are one-dimensional form.

STEADY-STATE MOLECULAR DIFFUSION

MASS TRANSFER

The basic equation governing mass transfer is Fick's first law

$$J_{A_y} = -D_{AB}\, \partial C_A / \partial y \tag{5.1}$$

where D_{AB} is the diffusivity of component A through B, and J_A is the molar flux in moles per unit time per unit area. Because mass flux is a vector,

$$J_A = -D_{AB}\,(\partial C_A / \partial x\ \underline{i} + \partial C_A / \partial y\ \underline{j} + \partial C_A / \partial z\ \underline{k}) \tag{5.2}$$

where $\underline{i}$, $\underline{j}$, and $\underline{k}$ are unit vectors in the x, y, and z direction, and J_A is the mass flux vector.

In defining J_A, it is stipulated that this flux must be referred to a plane across which there is no net volume transport (i.e., the plane moves with respect to the fixed apparatus although the fluid is stagnant; Figure 5.2). While this definition of J_A is physically correct, it is not useful in design and process work. What we need, instead, is a flux defined relative to the apparatus itself, rather than to a moving plane. We call this new flux N, which we shall derive.

The velocity of the moving plane can be given by

$$U_y = \sum_i N_i \bar{\nu}_i \tag{5.3}$$

where U_y is the velocity of the moving plane in distance per unit time, N_i is the mass flux of the ith component in moles per unit area per unit time, and $\bar{\nu}_i$ is the partial molal volume of the ith component in volume per mole.

For a binary system with components A and B

$$U_y = N_A \bar{\nu}_A + N_B \bar{\nu}_B \tag{5.4}$$

$$N_A = U_y C_A + J_A \tag{5.5}$$

and

$$N_A - C_A(N_A \bar{\nu}_A + N_B \bar{\nu}_B) = J_A = -D_{AB}\, \partial C_A / \partial y \tag{5.6}$$

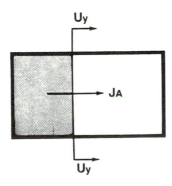

FIGURE 5.2. In this depiction of the motion of a diffusion boundary, flux J_A consists of the molecules that move across the moving boundary of the mass transferred. N_A combines convective mass transfer and molecular diffusion.

A mass balance on an overall basis is simple because outflow minus inflow equals accumulation or depletion. To consider a mass balance of individual species, we must treat each species separately with its own mass balance. A mass balance for a single species is called the equation of continuity of species.

To derive such a specific mass balance equation, we must consider the fluxes of mass in and out of a space element (Figure 5.3). We also have to consider chemical reaction because it can change one species to another.

The equation of continuity for a given species A in rectangular coordinates is

$$\partial C_A/\partial t + \partial N_{A_x}/\partial x + \partial N_{A_y}/\partial y + \partial N_{A_z}/\partial z = R_A \qquad (5.7)$$

where R_A is the rate of chemical reaction in moles per cubic centimeter-second.

For cylindrical coordinates, the mass balance is

$$\partial C_A/\partial t + 1/r\, \partial(rN_{A_r})/\partial r + 1/r(\partial N_A/\partial\theta) + \partial N_{A_z}/\partial z = [R_A] \qquad (5.8)$$

Equations 5.7 and 5.8, and their counterparts in spherical coordinates, represent the most important equations in chemical and petroleum processing, for they tie together chemical reaction and mass transfer.

Diffusivity, the transfer coefficient or property that relates flux and gradient, varies according to the combination of phases involved. Table 5.2 shows typical diffusivity values. The diffusivities of components in gases are orders of magnitude greater than their values in liquids or solids. These values point out that diffusion within a liquid or solid is much slower than the diffusion of a molecule at the interface of a liquid or solid with a gas.

Not shown in Table 5.2 is the effect of the diffusing species' concentrations on the diffusion coefficient. Concentration has a large effect on the diffusion coefficient in liquid and solid systems but not in gas systems at a given pressure regardless of

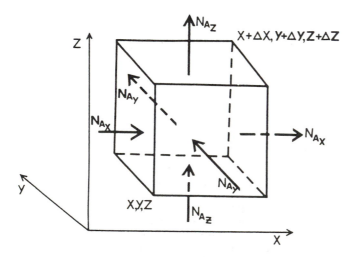

FIGURE 5.3. A mass balance on a volume in space accounts for mass flux into and out of the space in three coordinates plus chemical reaction.

TABLE 5.2. Typical Diffusivity Values for Various Systems

System	Diffusivity (cm^2/s)
Gas–gas	10^{-1}
Liquid–gas	10^{-6}
Gas–solid	10^{-7} to 10^{-10}
Solid–solid	10^{-15} to 10^{-30}

the amounts of A or B. (D_{AB} is the same value for gaseous A–B mixes whether the concentrations are 10–90, 30–70, 50–50, or 60–40.)

The process of mass transfer is analogous to that of heat transfer as we can see by comparing eq 4.6, the equation of energy

$$\partial/\partial x(k\partial T/\partial x) + \partial/\partial y(k\partial T/\partial y) + \partial/\partial z(k\partial T/\partial z) + \dot{q} = \rho C_p\, \partial T/\partial t$$

and eq 5.7, the equation of continuity of species

$$\partial C_A/\partial t + \partial N_{A_x}/\partial x + \partial N_{A_y}/\partial y + \partial N_{A_z}/\partial z = R_A$$

Both equations have accumulation terms ($\rho C_p\, \partial T/\partial t$ and $\partial C_A/\partial t$), generation terms ($\dot{q}$ and R_A), and flux terms (conduction and the N_A term). Temperature and concentration are analogous parameters.

Earlier we discussed heat transfer by considering steady-state heat conduction across a slab. Similarly, we begin considering mass transfer with steady-state molecular diffusion, illustrated in Example 5.1.

EXAMPLE 5.1. STEADY-STATE MOLECULAR DIFFUSION THROUGH A SLAB

Consider the steady-state diffusion of gaseous component A through a slab representing a gaseous mixture of A and B (Figure 5.4). (The slab represents a mixture unchanging with time.) What is the mass flux of component A through the slab in the y direction?

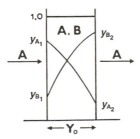

FIGURE 5.4.   This is a model for diffusion through a slab. Pure A is on the left side of the system. Molecules of A diffuse through the stagnant film of B ($y_2 - y_1$). B does not diffuse. (Reproduced with permission from reference 4. Copyright 1975 McGraw-Hill.)

With steady-state conditions ($\partial C_A/\partial t$ is zero) and no chemical reaction (R_A is zero), the equation of continuity of species becomes

$$\partial N_{A_y}/\partial y = 0$$

Because there is no flux of B, eq 5.6 becomes

$$N_A - C_A N_{A_y} \bar{v}_A = -D_{AB}\, \partial C_A/\partial y$$

Assuming that the gases are ideal, then

$$y_A = C_A/C_{total} = C_A RT/P$$

where y_A is the mole fraction of A, P is the total pressure, R is the gas constant, and T is the absolute temperature. Also

$$N_{A_y} = -D_{AB}P/RT(1 - y_A)\, dy_A/dy$$

Substitution $\partial N_{Ay}/\partial y = 0$ in the expression gives

$$d/dy\, [D_{AB}(dy_A/dy)/(1 - y_A)] = 0$$

Assuming D_{AB} is independent of concentration

$$d/dy\, [(dy_A/dy)/(1 - y_A)] = 0$$

Using the boundary conditions,

$$y = y \qquad y_A = y_A$$
$$y = y_2 \qquad y_A = y_{A_2}$$
$$(1 - y_A)/(1 - y_{A_1}) = [(1 - y_{A_2})/(1 - y_{A_1})]^{(y - y_1)/(y_2 - y_1)}$$
$$y_B/y_{B_1} = [y_{B_2}/y_{B_1}]^{(y - y_1)/(y_2 - y_1)}$$

Likewise, for flux N_{A_y}

$$N_{A_y}(1 - y_A) = [-D_{AB}P/RT]\, dy_A/dy$$

For the slab of thickness, y_0, integration gives

$$N_{A_y} = [D_{AB}P/RTy_0]\, \ln(y_{B_2}/y_{B_1})$$

Example 5.1 is applicable to a number of processes. We could place a graduated cylinder partially filled with an organic fluid in a laboratory hood. The air in the cylinder above the fluid plays the role of component B, a stagnant layer. It would be possible to estimate the diffusivity of the fluid vapor in air (using the amount evaporated, the fluid's vapor pressure, the temperature, and the total pressure). Other actual mass transfer cases can also be described by the stagnant layer approach.

UNSTEADY-STATE MOLECULAR DIFFUSION— TIME DEPENDENCE

In Chapter 4 we considered unsteady-state heat conduction. In mass transfer the parallel is unsteady-state molecular diffusion when mass transfer is a function of time.

To begin, we convert the equation of continuity of species, eq 5.7, to a form that changes N values to velocities and molecular diffusion. The result, for constant density and diffusivity, is

$$\frac{\partial C_A}{\partial t} + \left(\frac{V_X \partial C_A}{\partial x} + \frac{V_y \partial C_A}{\partial y} + \frac{V_z \partial C_A}{\partial z} \right) = D_{AB} \left(\frac{\partial^2 C_A}{\partial x^2} + \frac{\partial^2 C_A}{\partial y^2} + \frac{\partial^2 C_A}{\partial z^2} \right) + R_A \quad (5.9)$$

The various terms in eq 5.9 have special significance. The partial derivative with time is the unsteady-state effect. Convection is represented by the velocity terms. Molecular diffusion is the set of partial second derivatives multiplied by D_{AB}. The last term, R_A, is for chemical reaction.

Mass transfer in the y direction in a system with zero velocity (no flow) and no chemical reaction is given by

$$\frac{\partial C_A}{\partial t} = D_{AB} \left(\frac{\partial^2 C_A}{\partial y^2} \right) \quad (5.10)$$

or

$$\frac{\partial y_A}{\partial t} = D_{AB} \left(\frac{\partial^2 y_A}{\partial y^2} \right) \quad (5.11)$$

Equation 5.11 is the basic form of Fick's second law of diffusion for unsteady-state molecular diffusion in a given direction. Fick's second law describes many processes and is important in metallurgical systems.

Fick's second law can be extended to more dimensions as well as to chemical reactions. The differential equations for this extension are complex and beyond the scope of this text. Two texts (*3,4*) tabulate many solutions to both Fick's first and second laws. Although the material in reference 4 is written for heat conduction, it can be used for mass transfer because of the analogy, shown in the following equations, between the two processes.

$$\frac{\partial C_A}{\partial t} = D_{AB} \left(\frac{\partial^2 C_A}{\partial y^2} \right) \quad (5.10)$$

and

$$\frac{\partial T}{\partial t} = \left(\frac{k}{\rho C_p} \right) \left(\frac{\partial^2 T}{\partial y^2} \right) \quad (5.12)$$

The charts described earlier for heat transfer (*see* Chapter 4) can also be used for mass transfer. This transfer is done by substituting the appropriate concentration for the temperature and mass diffusivity D_{AB} for thermal diffusivity ($k/\rho C_p$). The analogy between heat and mass transfer follows a regular pattern:

Heat Transfer	Mass Transfer
Steady-state heat conduction	Steady-state molecular diffusion
Unsteady-state heat conduction	Unsteady-state molecular diffusion

MASS TRANSFER AT A PHASE BOUNDARY

The theoretical and basic equations for mass transfer can be solved for simple geometries and laminar flow where the flow field is specified. These cases can be mathematically complicated. If we consider turbulent flow and complex geometries, we cannot handle the situation analytically and classically simply or in some cases not at all. This situation clearly parallels that of heat transfer. Hence, we can define a mass transfer coefficient k and extend the analogy to heat transfer:

Heat Transfer	Mass Transfer
Steady-state heat conduction	Steady-state molecular diffusion
Unsteady-state heat conduction	Unsteady-state molecular diffusion
Convective heat transfer (heat transfer coefficients)	Convective mass transfer (mass transfer coefficients)

There are a number of interrelated mass transfer coefficients.

$$N_A = k_c(C_{A_1} - C_{A_2}) = k_y(y_{A_1} - y_{A_2}) = k^\star(y_{A_1} - y_{A_2})/y_{BM} = k_g(P_{A_1} - P_{A_2}) \quad (5.13)$$

The k_c, k_y, $k^\star$, and k_g are all mass transfer coefficients that, respectively, have units of centimeters per second, gram-moles per second-square centimeter, gram-moles per second-square centimeter, and gram-moles per second-square centimeters-atmosphere. The P_{AS} are partial pressures for component A, and y_{BM} is the logarithmic mole fraction of the nondiffusing component,

$$y_{BM} = (y_{A_1} - y_{A_2})/\ln (y_{B_2}/y_{B_1}) \quad (5.14)$$

The mass transfer coefficient, like its counterpart the heat transfer coefficient, is related to dimensionless groups. One of these groups is the Reynolds number. Another is a mass transfer group that plays the same role as the Prandtl number. This new group is defined as the Schmidt number,

$$\text{Schmidt number (Sc)} = \text{momentum diffusivity/mass diffusivity} \quad (5.15)$$

$$\text{Sc} = (\mu/\rho)/D_{AB} = \mu/\rho D_{AB} \quad (5.16)$$

A relationship for all of the mass transfer coefficient values for k is

$$k_c y_{BM} D/D_{AB} = k_y RT y_{BM} D/P D_{AB} = k^* RTD/P D_{AB} = f(\text{Re, Sc, } L/D) \quad (5.17)$$

Other k values can be defined using other driving forces, for example, k_ρ based on densities.

MASS TRANSFER FOR DIFFERENT GEOMETRIES

As mentioned earlier, the transfer of heat and mass are analogous for similar geometries, boundary conditions, and flow. For example [5],

$$j_D = j_H \quad (5.18)$$

where j_D, the mass transfer flux, is equal to j_H, the heat transfer flux,

$$j_D = k_c/V(\mu/\rho D_{AB})^{0.67} \quad (5.19)$$

and

$$j_H = h/C_p \rho V(C_p \mu/k)^{0.67} \quad (5.20)$$

This identity holds for flow over flat plates, cylinders, through packed beds, and in pipes and conduits (where Re $\geq$ 10,000).

Correlations exist for various geometries. The Reynolds number is correlated with the coefficients for flow through tubes (Figure 5.5). For flow over flat plates [6],

$$j_D = (k_c)_{average}(P_{BM}/VP)Sc^{0.67} \quad (5.21)$$

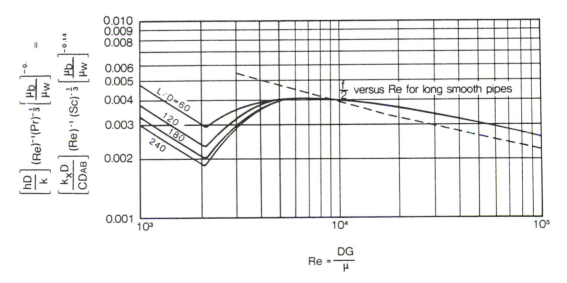

FIGURE 5.5. Heat and mass transfer are analogous for certain geometries. For flow through tubes, the Reynolds number Re can be correlated with both the heat transfer coefficient and the mass transfer coefficient, using the same data. (Adapted from reference 20.)

and (6)

$$j_D = j_H = f/2 = 0.037 \text{Re}^{-0.2} \tag{5.22}$$

where $8{,}000 < \text{Re} < 300{,}000$ and $\text{Re} = X_T V \rho / \mu$. X_T is the plate length, and P_{BM} is the partial pressure version of eq 5.14.

For a wetted wall column with Re based on gas velocity relative to the metal tube (7),

$$k_c D P_{BM}/D_{AB} P = 0.023 \text{Re}^{0.83} \text{Sc}^{0.44} \tag{5.23}$$

where $\text{Re} > 2000$.

If mass is transferred from a solid surface to a falling film, then the mass transfer coefficient is related to the film's physical characteristics (Figure 5.6). In the figure, Γ is the mass flow rate per unit perimeter or width, x is the vertical distance, ρ is the density, y_o is the film thickness, t is the residence time, and D_{AB} is the diffusivity.

For mass transfer between a fluid and a solid spherical particle, where D_p is the particle diameter and V_∞ is the particle's velocity of approach (8).

$$k_c D_p/D_{AB} = 2 + 0.60(D_p V_\infty \rho / \mu_f)^{0.5}(\rho / \mu D_{AB})^{0.33} \tag{5.24}$$

If flows are very low, then (9)

$$k_c D_p/D_{AB} = [4.0 + 1.21(\text{ReSc})^{0.67}]^{0.5} \tag{5.25}$$

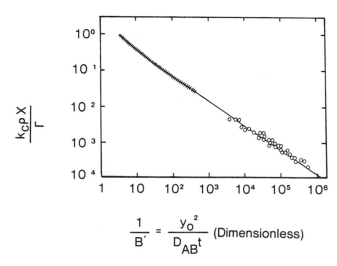

$$\frac{1}{B'} = \frac{y_o{}^2}{D_{AB} t} \quad \text{(Dimensionless)}$$

FIGURE 5.6. Mass transfer from a solid to a film falling over it is described by this correlation. In the figure, k_c is the mass transfer coefficient in centimeters per second, Γ is the mass flow rate per unit perimeter or width, X is the vertical distance that the film falls, ρ is the density of the liquid in the film, y_o is the film thickness, t is the residence time, and D_{AB} is the diffusivity. (Adapted from references 10, 21–23.)

Equation 5.25 provides a better description of the process in the lower Reynolds number range than does eq 5.24. When the Reynolds number is zero, both equations give the same result, namely, $k_c D_p / D_{AB}$ equals 2.0.

Figure 5.7 gives the analogy for heat and mass transfer for particles suspended in an agitated vessel (a highly complex system). For mass transfer from or to drops (*10*),

$$k_c D_p / D_{AB} = 1.13 \mathrm{Re}^{0.5} \mathrm{Sc}^{0.5} \tag{5.26}$$

If instead of drops, we have bubbles, then (*11*),

$$k_c D_p / D_{AB} = 1.13 (\mathrm{ReSc})^{0.5} [D_p / (0.45 + 0.2 D_p)] \tag{5.27}$$

where D_p is expressed in centimeters.

Rotating disks are devices that are widely used in electrochemical applications. The mass transfer Nusselt number for laminar flow over rotating disks, given a disk diameter D, is (*12*)

$$k_c D / D_{AB} = 0.879 \mathrm{Re}^{0.5} \mathrm{Sc}^{0.5} \tag{5.28}$$

For turbulent flow over disks (*13*),

$$k_c D / D_{AB} = 5.6 \mathrm{Re}^{1.1} \mathrm{Sc}^{0.33} \tag{5.29}$$

where $6 \times 10^5 < \mathrm{Re} < 10^6$, and $120 < \mathrm{Sc} < 1200$.

Additional correlations are shown in Figures 5.8 (cylinders) and 5.9 and 5.10 (packed beds).

The analogy between heat, mass, and momentum transfer is given for rotating cylinders (Figure 5.11). A typical application of the information in Figure 5.11 would be calculations for a rotating kiln or ore-roasting oven.

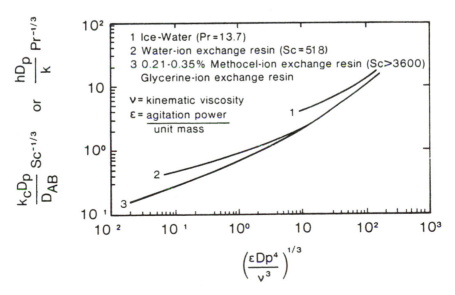

FIGURE 5.7.   This chart represents the heat and mass transfer correlation for particles suspended in an agitated vessel, where D_p is the particle diameter. (Adapted from references 9, 10, 24.)

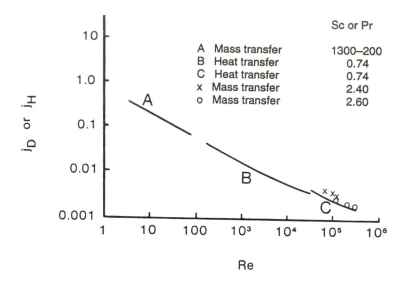

FIGURE 5.8. Heat and mass transfer can be correlated to Re for flow over cylinders that are normal to flow. j_D represents mass transfer, and j_H represents heat transfer. (Adapted from references 10, 25–29.)

Processes involving convective mass transfer are handled in a manner similar to heat transfer cases:

1. Find an average temperature.
2. Determine the fluid properties, such as density or viscosity at the average temperature.
3. Calculate the Reynolds and Schmidt numbers.
4. Select the appropriate convective equation.
5. Calculate the required information.

We shall consider some pertinent examples.

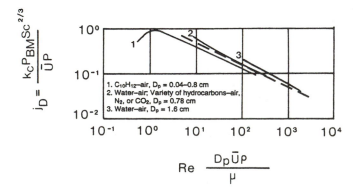

FIGURE 5.9. The correlation between flow and mass transfer for a gas flowing through a bed of solid particles. $\bar{U}$ is the superficial velocity (the velocity in the unit if the packing were not present). (Adapted from references 10, 30–34.)

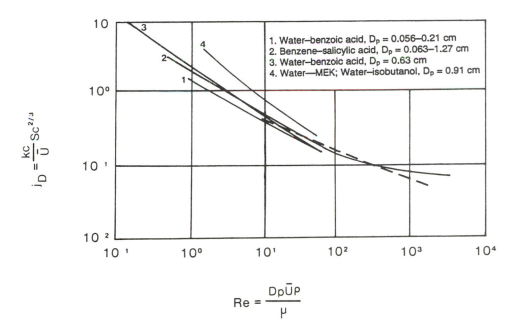

FIGURE 5.10. The correlation between flow and mass transfer for a liquid flowing through a bed of solid particles. (Adapted from references 10, 35–38.)

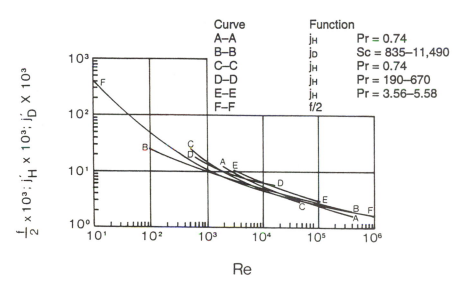

FIGURE 5.11. Heat, mass, and momentum transfer have been correlated with Re for rotating cylinders contacting liquids or gases. (Adapted from references 10, 39–44.)

EXAMPLE 5.2. SUBLIMATION RATE FROM A CYLINDER

Air passes through a naphthalene tube that has an inside diameter of 0.0254 m and a length of 1.83 m. The velocity is 15.24 m/s and the air is at 10 °C and atmospheric pressure. What is the percent saturation of air with naphthalene, and what is the rate of naphthalene sublimation at 15.24 m/s given air properties:

- $\rho = 1.249$ kg/m^3 and
- $\mu = 1.8 \times 10^{-5}$ kg/m-s

and naphthalene properties:

- vapor pressure = 2.79 N/m^2,
- $D_{AB} = 5.2 \times 10^{-6}$ m^2/s, and
- molecular weight = 128.2.

The Reynolds number is

$$\text{Re} = (0.0254 \text{ m})(15.24 \text{ m/s})(1.249 \text{ kg/m}^3)/(1.8 \times 10^{-5} \text{ kg/m-s})$$

$$\text{Re} = 26{,}860$$

The Schmidt number is

$$\text{Sc} = (1.8 \times 10^{-5} \text{ kg/m-s})/(1.249 \text{ kg/m}^3)(5.2 \times 10^{-6} \text{ m}^2/\text{s})$$

$$\text{Sc} = 2.77$$

Then, let us use a k_ρ

$$k_\rho = k_c \left(P_{BM}/P\right)$$

and

$$k_\rho D/D_{AB} = k_c D P_{BM}/D_{AB} P = 0.023 \ \text{Re}^{0.8}\text{Sc}^{0.33}$$

$$k_\rho = 0.0229 \text{ m/s}$$

To deal with the mass transferred, we need a balance analogous to $Q = \Delta H$ in heat transfer. Our balance equates the mass taken up by the air and the mass transferred. We also use differential changes for the length of the tube (dx) and the change of density of the napthalene vapor ($d\rho_A$). The $(\rho_{A \text{ sat}} - \rho_{A \text{ bulk}})$ is the driving force for mass transfer. Hence, the mass taken up by the air equals the mass transferred, so

$$(\pi/4)(0.0254 \text{ m})^2(15.24 \text{ m/s})d\rho_{A \text{ bulk}} = k_\rho \pi(0.0254 \text{ m})dx(\rho_{A \text{ sat}} - \rho_{A \text{ bulk}})$$

Then,

$$\int_0^{\rho_{A \text{ sat}}} d[\rho_{A \text{ bulk}}/(\rho_{A \text{ sat}} - \rho_{A \text{ bulk}})] = 2.58 k_\rho \int_0^{1.83} dx$$

$$-\ln\left[(\rho_{A \text{ sat}} - \rho_{A \text{ bulk}})/\rho_{A \text{ sat}}\right] = 0.43$$

and

$$\rho_{A \text{ bulk}} = 0.35 \ \rho_{A \text{ sat}}$$

Hence, saturation is 35%.

At 10 °C, saturation of naphthalene in air is 1.52×10^{-4} kg/m^3 because

$$\rho_{A \text{ sat}} = (2.79 \text{ N/m}^2)(128)(1.249 \text{ kg/m}^3)/(1.01 \times 10^5 \text{ N/m}^2)(29)$$

Then,

total evaporation rate =

$$(\pi/4)(0.0254 \text{ m})^2(15.24 \text{ m/s})(0.35)(1.52 \times 10^{-4} \text{ kg/m}^3)$$

$$\text{total evaporation rate} = 4.2 \times 10^{-7} \text{ kg/s}$$

This last example has some points that deserve further discussion. The first point is the use of the mass transfer coefficient k_ρ based on density, which was not previously mentioned. Saturation calculations, such as for humidity, are based on densities. Also, using k_ρ means that P_{BM}, a term that requires the partial pressures of air at various points for calculations, did not have to be used.

A second point is that when calculating Sc, the diffusivity value was that for naphthalene in air, D_{BA}, whereas the other values such as for density and viscosity were those for air. The use of these values appears to be inconsistent. However, the diffusivities of naphthalene in air and air in naphthalene are identical. An explanation for this can be derived from the kinetic theory of gases, which yields identical expressions for D_{AB} and D_{BA}. (In brief, the diffusivities are derived from a collection of the same terms involving temperature, pressure, and other parameters multiplying either $[(1/M_A) + (1/M_B)]^{0.5}$ for D_{AB} or $[(1/M_B) + (1/M_A)]^{0.5}$ for D_{BA}; M_A and M_B are the molecular weights of components A and B, respectively. Hence, $D_{AB} = D_{BA}$ in a binary system.) The relationship D_{AB} equals D_{BA} holds for binary mixtures, but it does not hold for multicomponent systems.

EXAMPLE 5.3. EVAPORATION RATE FROM A DROPLET

A spherical drop of water (0.05 cm in diameter) is falling at a velocity of 215 cm/s through dry, still air at 1 atmosphere. Estimate the instantaneous rate of evaporation from the drop if the drop surface is at 70 °F and the air is at 140 °F.

To solve this problem, we shall assume ideal gas behavior, insolubility of air in water, equilibrium at the interface, and pseudosteady-state conditions.

For a small evaporation rate, we can write a mass transfer coefficient equivalent to eq. 5.6,

$$W_A - X_{A0}(W_A + W_B) = k^*A(X_{A0} - X_{A\infty})$$

where W_A is the molar rate of exchange of A, and X_{A0} and $X_{A\infty}$ are the water mole fractions at the drop surface and in the air beyond the boundary layer around the drop.

$$W_A = k^*A(X_{A0} - X_{A\infty})/(1 - X_{A0})$$
$$W_A = k^* \pi D^2(X_{A0} - X_{A\infty})/(1 - X_{A0})$$
$$T_f = (70 + 140)/2 = 105 \text{ °F}$$
$$X_{A0} = 0.0247 \text{ atm}/1 \text{ atm} = 0.0247$$

and

$$X_{A_\infty} = 0$$

Also,

$$C_f = 3.88 \times 10^{-5} \text{ g-mol/cm}^3$$
$$\rho_f = C_f M = 1.12 \times 10^{-3} \text{ g/cm}^3$$
$$\mu_f = 1.91 \times 10^{-4} \text{ g/cm-s}$$
$$D_{AB} = 0.292 \text{ cm}^2/\text{s}$$

where the subscript f indicates that properties were evaluated at T_f.

$$\text{Sc} = (\mu_f/\rho_f D_{AB}) =$$
$$(1.91 \times 10^{-4} \text{ g/cm-s})/(1.12 \times 10^{-3} \text{ g/cm}^3)(0.292 \text{ cm}^2/\text{s})$$
$$\text{Sc} = 0.58$$

and

$$\text{Re} = (DV\rho_f/\mu_f)$$

$$\text{Re} = (0.05 \text{ cm})(215 \text{ cm/s})(1.12 \times 10^{-3} \text{ g/cm}^3)/(1.91 \times 10^{-4} \text{ g/cm-s})$$
$$\text{Re} = 63$$

Using eq 5.24,

$$k^* = C_f D_{AB}/D_p \, [2 + 0.60(D_p V_\infty \rho/\mu f)^{0.5}(\rho/\mu D_{AB})^{0.33}]$$
$$k^* = (3.88 \times 10^{-5} \text{ g-mol/cm}^3)(0.292 \text{ cm}^2/\text{s})/$$
$$(0.05 \text{ cm}) \, [2 + 0.60(63)^{0.5}(0.58)^{0.33}]$$
$$k^* = 1.35 \times 10^{-3} \text{ g-mol/s-cm}^2$$

Then,

$$W_A = k^* \pi D^2 (X_{A0} - X_{A_\infty})/(1 - X_{A0})$$
$$W_A = (1.35 \times 10^{-3} \text{ g-mol/s-cm}^2)\pi(0.05 \text{ cm})^2[(0.0247 - 0)/1 - 0.0247)]$$
$$W_A = 2.70 \times 10^{-7} \text{ g-mol/s}$$

This molar flow rate amounts to a decrease of 1.23×10^{-3} cm/s in the drop diameter. Hence, a drop would fall a considerable distance before evaporating.

   If we were to do Example 5.3 on a rigorous basis, not making the assumptions we did, we would combine heat and mass transfer, and also the change in drop diameter as the water evaporates. This change would require a differential equation. However, the simpler approach yielded acceptable results.

EQUILIBRIUM STAGE OPERATIONS, THE IDEAL STAGE, AND PHASE EQUILIBRIA

Diffusional mass transfer describes many physical processes. However, it is difficult in many instances to use this technique to design large-scale equipment. For such cases, we use an approach based on an ideal stage (Figure 5.12). Its contents are so well mixed that streams leaving it are in equilibrium. In an ideal stage, the mass transfer takes place because the entering streams, which are not at equilibrium, reach equilibrium. The differences in composition between the entering and equilibrium values constitute the driving forces.

Mass transfer operations such as distillation, absorption, extraction, leaching, and crystallization can be analyzed by the ideal stage concept. Solutions for these mass transfer processes or the design of devices to carry out such operations require calculations of (1) a material or mass balance, (2) an enthalpy or energy balance, and (3) equilibrium data. The first two items are the conservation of mass and conservation of energy. The third item represents the appropriate relationship between the phases present, for example, vapor–liquid equilibrium data for distillation. Table 5.3 shows the type of data required for a given process (*see also* Figures 5.13–5.16).

Figures 5.14 and 5.15 both represent the vapor–liquid equilibrium between benzene and toluene. Their interchangeability can be explained by the use of the Gibbs phase rule

$$P + V = C + 2$$

where P is the number of phases, V represents the variables or degrees of freedom, and C is the number of components. There are two phases and two components in the benzene–toluene system. Hence, there are two degrees of freedom. One of these degrees is specified by pressure, which leaves just one variable. If the benzene mole fraction in either liquid or vapor is specified, then we automatically fix the temperature and mole fraction in the other phase. The reader should verify this exercise by studying Figures 5.14 and 5.15.

MASS BALANCES FOR AN IDEAL STAGE

The concept of an ideal stage can be explained by examining a continuous distillation column (Figure 5.17). Calculations for a mass balance across a continuous distillation are simpler than for a batch distillation. Once the continuous column reaches

FIGURE 5.12. Calculations for mass transfer operations, such as distillation and absorption, use the concept of the process as a series of ideal stages. An ideal or equilibrium stage is so well-mixed that the exiting streams, D and C in this illustration, are in equilibrium. Nonequilibrium streams, A and B, provide the driving force for mass transfer within the stage.

TABLE 5.3. Types of Equilibrium Data Needed To Calculate the Mass Transfer in Ideal Stage Operations

Process Types	Data Required
Distillation	Vapor–liquid
Extraction	Liquid–liquid
Absorption	Vapor–liquid
Leaching	Solid–liquid
Adsorption	Solid–liquid or solid–gas
Crystallization	Solid–liquid

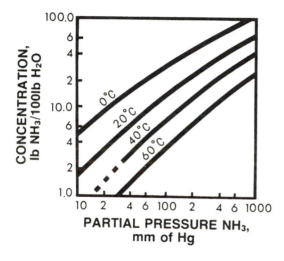

FIGURE 5.13. Equilibrium calculations for an ideal stage in mass transfer equipment, such as an absorption column, may require solubility data. These data are for the solubility of ammonia in water. (Reproduced with permission from reference 46. Copyright 1967 McGraw-Hill.)

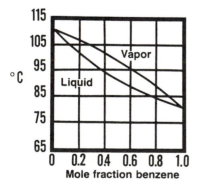

FIGURE 5.14. Boiling point data, such as this diagram for benzene and toluene, is helpful when calculating the number of ideal stages needed for a distillation column. (Reproduced with permission from reference 46. Copyright 1967 McGraw-Hill.)

its steady state, all of the streams' concentrations remain constant with time. Batch columns, however, have concentrations that change with time as the volatile materials first distill, followed by less volatile components. Illustrations of distillation columns, packing materials, and packed columns are given (Figures 5.17–5.21).

In setting up the mass balances for the column shown in Figure 5.17, we will

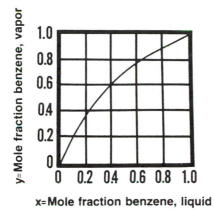

FIGURE 5.15. If ideal stage calculations are performed for a two-component system, the equilibrium data needed may be a vapor–liquid mole fraction diagram, such as this one for benzene and toluene. (Reproduced with permission from reference 46. Copyright 1967 McGraw-Hill.)

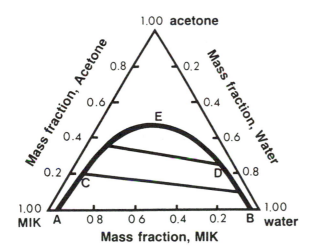

FIGURE 5.16. Reading a ternary diagram. Separation processes sometimes make use of an extracting agent in which the solute is soluble and the solvent is not. The three components form a ternary system. This diagram is for a ternary system of water, MIK (methyl isobutyl ketone), and acetone. Each corner of the triangle represents 100% of a component. Each side opposite a point represents 0% of the component. The area under the curved line ACEDB is formed of a mixture of two saturated and insoluble liquid phases, the composition of each defined by the points at the end of tie lines crossing the area. In such a ternary mixture, the closer points A and B come to the points of the triangle, the more insoluble are the liquids A and B in each other. The driving force behind the extraction at each ideal stage is the difference in the solubility of acetone in A-rich and B-rich phases related by a tie line (see Example 5.5). (Adapted from reference 47.)

only consider the streams cut by the given boundary. On an overall basis, we use the AAAA boundary. The overall mass balance for total streams is

$$F = D + B \tag{5.30}$$

where F, D, and B are the moles per hour of the feed, overhead (distillate) product, and the bottoms product stream, respectively. The mass balance on the most volatile stream would be

$$FX_F = DX_D + BX_B \tag{5.31}$$

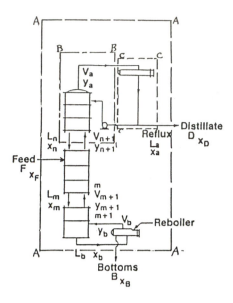

FIGURE 5.17. A distillation column can be examined from several mass balance points of view. A balance for the entire column would be defined by boundary AAAA, the enriching section above the feed point by boundary BBBB, and the reflux section by boundary CCCC. F is the feed, D is the distillate, and B is the bottoms product, all in moles per hour. X is the mole fraction of component in the liquid, and Y is the fraction in the vapor. (Adapted from reference 48.)

where X represents the fraction of component A in the liquid feed, distillate, and bottoms streams.

Solving eqs 5.30 and 5.31 for the ratios of distillate and bottoms to feed results in

$$D/F = (X_F - X_B)/(X_D - X_B) \qquad (5.32)$$

and

$$B/F = (X_D - X_F)/(X_D - X_B) \qquad (5.33)$$

Next we will look at the enriching or rectifying section of the column (the sections above the feed in Figure 5.17) with boundaries BBBB and CCCC in which n is the stage number, Y is the mole fraction of a in the vapor stream, V is the molar flow rate of the vapor, and L is the molar flow rate of the liquid. The boundary BBBB gives

$$(V_{n+1})(Y_{n+1}) + (L_a X_a) = (V_a Y_a) + (L_n X_n) \qquad (5.34)$$

and the boundary CCCC yields

$$DX_D + (L_a X_a) = (V_a Y_a) \qquad (5.35)$$

Combining the two equations yields

$$Y_{n+1} = L_n X_n/V_{n+1} + DX_D/V_{n+1} \qquad (5.36)$$

Finally, a boundary (not shown in Figure 5.17) around the entire portion of the column and reflux unit that cuts L_n, V_{n+1}, and D, shows that V_{n+1} equals $L_n + D$. Substitution in eq 5.36 gives

$$Y_{n+1} = L_n X_n/(L_n + D) + DX_D/(L_n + D) \qquad (5.37)$$

a Bubble cap system

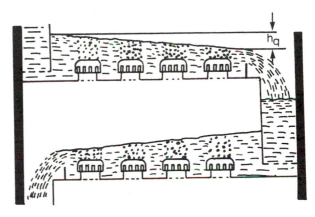

b Detail of bubble cap

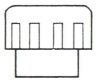

c Sieve plate system

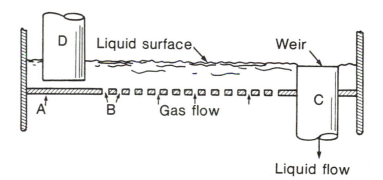

FIGURE 5.18. A stage can take various mechanical forms in process equipment. Process units, such as a distillation column in a petrochemical plant, can be more than 100 ft tall with hundreds of stages (trays). (a) In a bubble tray, each stage is composed of a perforated sheet of metal with a weir cutting across one side and a bubble cap covering each perforation. The cap (b) has slits through which rising vapor passes, bubbling through the downflowing liquid. (Parts a and b are reproduced with permission from reference 49. Copyright 1968 Pergamon.) In a sieve tray (c), gas flows upward through holes in the tray and contacts the liquid flowing over the tray to the weir. (Part c is reproduced with permission from reference 46. Copyright 1967 McGraw-Hill.)

Raschig ring   **Lessing ring**   **Berl saddle**   **Intalox saddle**

Tellerette   **Pall ring**

FIGURE 5.19.   Contact between vapor and gas in separation processes, such as extraction or distillation columns, can be enhanced by filling the columns with packing material. Column packing comes in many shapes, sizes, and materials. (Reproduced with permission from reference 10. Copyright 1975 McGraw-Hill.)

A similar approach for the stripping section (the stages below the feed) gives eq 5.38 and 5.39. To understand the mass balance process, it is best to verify these equations by working them out in a manner similar to that used for eq 5.36 and 5.37.

$$Y_{m+1} = (L_m/V_{m+1})X_m - (B/V_{m+1})X_B \tag{5.38}$$

and

$$Y_{m+1} = [L_m/(L_m - B)]X_m - [B/(L_m - B)]X_B \tag{5.39}$$

Liquid

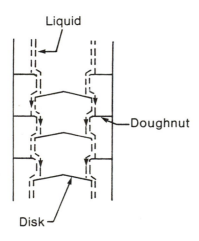

Doughnut

Disk

FIGURE 5.20.   In a disk-and-doughnut column, vapor is forced to contact the down-flowing liquid by being diverted to the edge by a disk. (Reproduced with permission from reference 50. Copyright 1978 Prentice-Hall.)

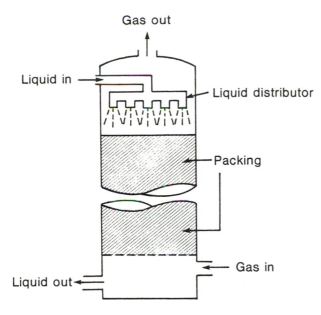

FIGURE 5.21. In a packed separation column, liquid is fed into the top of the column and flows down over the packing, while gas flows upward. (Reproduced with permission from reference 50. Copyright 1978 Prentice-Hall.)

ENERGY BALANCE FOR AN IDEAL STAGE

To balance the energy across a stage, we use a base enthalpy of zero for a liquid (L_n) at T_n and the types of data specified in Table 5.4, given stage n at temperature T_n. The energy balance is then

$$a + b - c - d + e - f = 0 \qquad (5.40)$$

In eq 5.40, a and d are much larger than the other terms, and the remaining terms are about the same order of magnitude. Hence we can neglect terms b, e, c, and f, leaving

$$a = d \qquad (5.41)$$

TABLE 5.4. Energy Values Needed for an Ideal Stage Balance

Item	Symbol
Latent heat in vapor V_{n+1}	a
Sensible heat in vapor above T_n	b
Sensible heat in liquid L_{n-1} below T_n	$-c$
Latent heat in vapor V_n	d
Heat of mixing	e
Convection and radiation loss	f

In turn,

$$V_n \lambda_n = V_{n+1} \lambda_{n+1} \tag{5.42}$$

For chemically similar materials,

$$V_n = V_{n+1} \tag{5.43}$$

If eq 5.36 holds,

$$L_{n-1} = L_n \tag{5.44}$$

and

$$L_{m-1} = L_m$$

THE McCABE–THIELE METHOD

When eqs 5.41–5.44 hold, the conditions are known as constant molal vaporization and constant molal overflow. These conditions mean that the liquid and vapor flow rates in the enriching and stripping sections are, respectively, equal. Therefore, eqs 5.36–5.39 are straight lines.

This method, known as the McCabe–Thiele method, is used to solve the relationship graphically between material balances and equilibrium for constant molal overflow and to calculate the number of stages in mass transfer equipment, such as distillation columns. A typical McCabe–Thiele solution is shown in Figure 5.22, in which each step represents a stage.

The starting point in a McCabe–Thiele diagram is an equilibrium diagram of the type shown in Figure 5.15. The first step is to depict the appropriate material balances on the graph.

The rectifying operating line (the upper line in Figure 5.22) can be placed. It must pass through the point $X_D = Y_D$ because at the top of the column (Figure 5.17), $Y_A = X_D = X_A$, and it intercepts the y axis at $X_D/(R_D + 1)$ where R_D is the reflux ratio, L_a/D. The stripping line (the lower line in Figure 5.22) must intercept both the feed line and the rectifying line. The feed line slope depends on the quality of the feed, whether it is a saturated liquid, super-heated vapor, etc. (Figure 5.23). The equation of the feed line is

$$Y = -(1 - f/f)X + X_F/f \tag{5.45}$$

where f is the feed mole condition. Values of f for various conditions are shown in Table 5.5.

Once the rectifying and stripping operating lines are correctly placed, we can determine the number of stages (the steps in Figure 5.22) required for separation. To create the first step, we start from the point $X_D = Y_D$ on the line $X = Y$, and move horizontally left to the equilibrium line, then vertically down to the rectifying operating line. (This sequence actually involves using the material balance of the rectifying operating line with the equilibrium data.) The step procedure continues until the feed point is reached. At this juncture, the material balance switches to the stripping line. Steps are continued until the X_B intercept is reached.

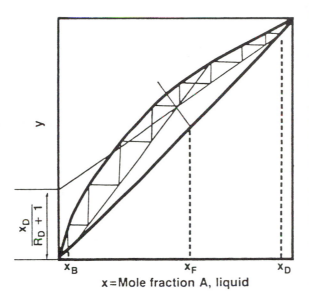

FIGURE 5.22. The McCabe–
Thiele method graphically determines the number of stages (the
steps drawn in this diagram) needed
for a separation by determining the
relationship between a material balance and vapor–liquid equilibrium
for one component, as shown here.
The solution starts with an equilibrium diagram, such as Figure 5.15,
which is the top limit for the steps.
The feed composition X_F is at the
junction of two mass balance lines,
one for the rectifying section above
the feed point of the column and
the other for the stripping section
below the feed point. The rectifying
line intersects the point $X_D = Y_D$
(the mole fraction of A in the liquid
and in the vapor of the distillate D)
and the y axis at a point determined
by the reflux ratio R_D and the composition of the distillate ($X_D/(R_D + 1)$).
The lower line, limiting the stages, is
the stripping line, which must intersect the rectifying and feed lines and
$X_B = Y_B$ for the bottoms B product.
(Adapted from reference 46.)

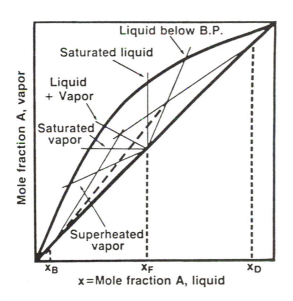

FIGURE 5.23. The equation
that defines the feed line (eq
5.45, $Y = [-(1 - f)/f]x +
X_F/f$) is determined by the con-
dition of the feed. Feed may be
cold, saturated liquid, saturated
vapor, or superheated vapor
(see Table 5.5 for values of f).
(Adapted from reference 51.)

EXAMPLE 5.4. DESIGNING A FRACTIONATING COLUMN WITH DIFFERENT FEED CONDITIONS

We need to design a continuous fractionating distillation column to separate 3.78
kg/s of 40% benzene and 60% toluene into an overhead product containing 97 mass
percent benzene and a bottom product containing 98 mass percent toluene. We must

TABLE 5.5. Feed Values for Varying Feed Conditions

Feed Condition	f Value
Cold feed	$f < 0$ and $f = -C_{liquid}(T_b - T_f)/\lambda$
Saturated liquid (bubble point)	$f =$ fraction vapor
Saturated vapor	$f = 1.0$
Superheated vapor	$f = 1 + C_{vapor}(T_f - T_d)/\lambda$

NOTE: T_b and T_d are the bubble and dew points for the feed.

use a reflux ratio of 3.5 mol to 1 mol of product. The latent molal heat of both benzene and toluene is about 357.1 kJ/kg. We will consider the following cases for the design: (1) liquid feed at its boiling point, (2) liquid feed at 20 °C ($C_{liquid} = 1.75$ kJ/kg-°C), and (3) feed is two-thirds vapor.

First, we determine the necessary material balances, mole fractions, etc.

$$X_F = (40/78)/(40/78 + 60/92) = 0.44$$
$$X_D = (97/78)/(97/78 + 3/92) = 0.974$$

and

$$X_B = (2/78)/(2/78 + 98/92) = 0.0235$$

The feed rate F is

$$F = (3.78/100 \text{ kg/s})/(40/78 + 60/92) = 0.0441 \text{ kg-mol/s}$$
$$D = 0.0441(X_F - X_B)/(X_D - X_B) = 0.0193 \text{ kg-mol/s}$$

and

$$B = F - D = (0.0441 - 0.0193) \text{ kg-mol/s} = 0.0248 \text{ kg-mol/s}$$

Now the McCabe–Thiele diagram can be drawn. First, we plot the equilibrium line and $X = Y$ line together with the vertical X_B, X_F, and X_D (Figure 5.24). Next we locate the rectifying operating line (passes through $X_D = Y_D$ and the y axis intercept of $X_D/(R_D + 1)$ or $0.974/(3.5 + 1) = 0.216$).

The f value for case 1 (liquid feed at the boiling point) is zero, giving a vertical feed line. The intercept of the feed line and the rectifying operating line gives one intercept for the stripping operating line (the other intercept is $X_B = Y_B$).

For case 2, f is given by

$$f = -C_{liquid}(T_b - T_f)/\lambda$$
$$f = -1.75(95 - 20)/357.1$$

and

$$f = -0.370$$

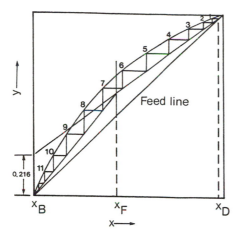

FIGURE 5.24. Example 5.4 uses a McCabe–Thiele diagram to determine the number of stages needed to separate benzene and toluene, given different feed scenarios. This diagram solves for the feed at its boiling point. (Reproduced with permission from reference 48. Copyright 1985 McGraw-Hill.)

The McCabe–Thiele diagram for case 2 (feed line slope of 3.70) is shown in Figure 5.25.

Finally, for case 3, the f value is $2/3$ (because $2/3$ of the feed is vapor), thus giving a slope of $-1/2$ for the feed line. The resultant solution is given in Figure 5.26. There are eleven ideal stages plus a reboiler with feed entering on the seventh plate from the top for cases 1 and 3. Case 2 gives 9 ideal stages plus reboiler with feed on the fifth stage from the top.

MINIMUM AND MAXIMUM STAGES IN AN IDEAL STAGE SEPARATION

The two limiting conditions in distillation are the minimum number of plates (total reflux) and the minimum reflux (infinite number of plates). The minimum number of plates or stages can be determined (Figure 5.27) by using the line $x = y$ as the operating line for both rectifying and stripping. Minimum reflux can be determined graphically (Figure 5.28). When one or both operating lines touch the equilibrium line, an infinite number of stages will result because we cannot infinitely subdivide matter. Understanding how to find the stages needed for minimum reflux is impor-

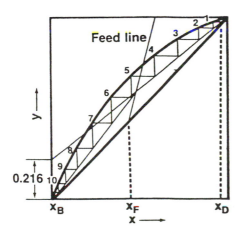

FIGURE 5.25. In Example 5.4, the McCabe–Thiele diagram for feed at 20 °C. (Reproduced with permission from reference 48. Copyright 1985 McGraw-Hill.)

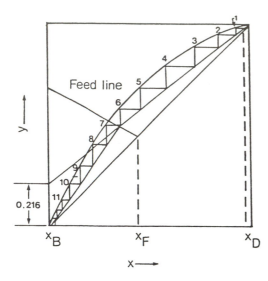

FIGURE 5.26. In Example 5.4, the McCabe–Thiele diagram for feed that is two-thirds vapor. (Reproduced with permission from reference 48. Copyright 1985 McGraw-Hill.)

tant because the actual reflux ratio used in distillation columns is 1.3 to 1.7 times the minimum (a figure of 1.5 is frequently used).

RELATIVE VOLATILITY—USING ENTHALPY–CONCENTRATION DIAGRAMS

A useful quantity in describing distillation is the relative volatility α_{AB} defined as

$$\alpha_{AB} = (Y_A/X_A)(X_B/Y_B) = Y_A(1 - X_A)/X_A(1 - Y_A) \qquad (5.46)$$

This equation for relative volatility indicates the ease or difficulty of achieving separation of A and B (large or small α values indicate easy separation while a value of 1 means A and B are inseparable). It also provides a means of generating equilibrium data because α is a fairly constant quantity at a given total pressure.

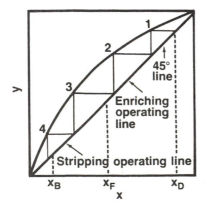

FIGURE 5.27. If the $x = y$ line of an equilibrium diagram is used as the lower limits of a McCabe–Thiele solution for both the rectifying and stripping sections (the condition of total reflux), the number of stages will be minimum. (Adapted from reference 46.)

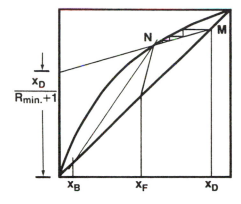

FIGURE 5.28. If the reflux is minimum, the number of stages needed for separation will be infinite, according to a McCabe–Thiele diagram.

THE PONCHON–SAVARIT METHOD OF NONCONSTANT MOLAL MASS TRANSFER

The McCabe–Thiele method relies on the assumptions of constant molal vaporization and constant molal overflow. When we cannot assume constant molal overflow (i.e., when there are significant thermodynamic nonidealities in the system), the overall enthalpy balance (eq 5.40) is important for determining the number of stages needed in separation equipment. In such cases, a graphical solution is found by using enthalpy–concentration diagrams that include equilibrium lines (Figure 5.29); this is called the Ponchon–Savarit method. Descriptions of the techniques can be found in a number of sources (*6,8,14*); the method is described briefly here and a full description is provided in reference 6:

1. Start with an enthalpy–concentration diagram (Figure 5.29), with saturated vapor and liquid lines (the bold curves).

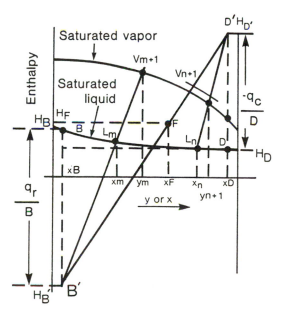

FIGURE 5.29. When constant molal overflow in a separation process cannot be assumed (as in the McCabe–Thiele method) and the energy balance becomes necessary, the number of stages needed can be determined by the Ponchon–Savarit method. The method is complex and is not covered in detail in this text. For further information, see references 6, 8, and 14. (Reproduced with permission from reference 46. Copyright 1967 McGraw-Hill.)

2. Assume that the distillate and bottoms products (D and B) are at bubble temperatures that fall on the saturated liquid line.
3. The feed F can be at any thermal condition; it is shown in Figure 5.29 as a mixture of liquid and vapor, with enthalpy, H_F, and concentration, x_F.
4. The process is nonadiabatic; for an adiabatic condition on this figure, product streams D and B must be corrected for the heat removed by the condenser ($-q_c$) and the heat added by the reboiler (q_r):

$$H'_D = H_D - (-q_c/D) = H_D + q_c/D$$

$$H'_B = H_B - q_r/B$$

In the diagram, points D' and B' represent corrected product streams. Line B'FD' is the overall enthalpy line.

5. Either $-q_c$ (usually the case) or q_r is taken to be independent; $-q_c$ is fixed by the reflux ratio

$$(H'_D - H_{y_1})/(H_{y_1} - H_D)$$

for which H_{y_1} is the specific enthalpy of vapor leaving the first stage.

6. An enthalpy operating line can then be constructed for each plate: in Figure 5.29, the line $L_n V_{n+1}$ describes the transfer taking place between plates n and $n+1$, where L_n is the liquid going from plate n to plate $n+1$, and V_{n+1} is the vapor leaving plate $n+1$ to go to plate n.
7. In a stepwise procedure, the number of ideal plates can be determined.

ESTIMATING THE STAGES NEEDED FOR A MULTICOMPONENT SEPARATION WITH RELATIVE VOLATILITIES

The design of a distillation column becomes complex when the system involves separation of more than two components. Separation of more than two components requires stage-by-stage calculations that carefully balance mass and energy, a laborious, time-consuming task best handled by a computer. There are methods, however, of estimating the number of stages needed for a given multicomponent system.

The first step in the process is to use the relative volatilities of the components to evaluate the system. Each volatility is considered relative to the least volatile component. For example, consider a system of five components (Table 5.6).

TABLE 5.6. Relative Volatilities for a System with More Than Two Components

Component	Volatility Relative to Least Volatile Component
A	5.1
B	3.6
C	1.7
D	1.3
E	1.0

The components closest to the point of separation are B and C; most of B is in the overhead stream and a little in bottoms and vice versa for C. This means that B is the *light key* (lk) and C is the *heavy key* (hk).

To separate a multicomponent mixture such as that as shown in Table 5.6, the minimum steps or stages are given (*15*) by

$$S_{\min} = \log \left[(X_{lk}/X_{hk})_D (X_{hk}/X_{lk})_W \right]/\log \alpha' \qquad (5.47)$$

where X represents the desired mole fractions of the heavy or light keys in the distillate (D) and bottom (W) products. The α' is the volatility of the light key relative to the heavy key (in the case cited above α' equals 3.6/1.7). Just as we assume for binary mixtures, a minimum number of stages assumes total reflux.

Next, for minimum reflux (*14*),

$$R_{\min} + 1 = \alpha_A X_{D_A}/(\alpha_A - \theta) + \alpha_B X_{D_B}/(\alpha_B - \theta) + \dots \alpha_Z X_{D_Z}/(\alpha_Z - \theta) \quad (5.48)$$

The αs are volatilities relative to the least volatile, the X_Ds are the mole fractions in the overhead product, and θ is an empirical constant. The constant θ is obtained by trial and error (*14*) from

$$f = \alpha_A X_{F_A}/(\alpha_A - \theta) + \alpha_B X_{F_B}/(\alpha_B - \theta) + \dots + \alpha_Z X_{F_Z}/(\alpha_Z - \theta) \quad (5.49)$$

where the f is the feed condition and the X_Fs are the mole fractions of the components in the feed. Again minimum reflux, just as with binary mixtures implies an infinite number of stages.

Once both $S_{\min}$ and $R_{\min}$ are known, we can determine the number of stages needed for any reflux ratio using a graphical correlation (Figure 5.30).

DIFFERENTIAL DISTILLATION FOR A BATCH DISTILLATION

The quantitative treatment of batch distillation is difficult and complex because the distillate composition changes with time. One aspect of these distillations that can be treated more simply is that of *differential distillation*. In differential distillation, the liquid is vaporized and each segment of vapor is removed from contact with the liquid as it is formed. Although the vapor can be in equilibrium with the liquid as it is formed, the *average vapor formed* will not be in equilibrium with the liquid residue.

For an amount of liquid $-dW$ to be vaporized,

$$-y\, dW = -d\,(Wx) \qquad (5.50)$$

and

$$W\, dx/dW = y - x \qquad (5.51)$$

so that

$$\int_{W_0}^{W} dW/W = \int_{x_0}^{x} dx/(y - x) \qquad (5.52)$$

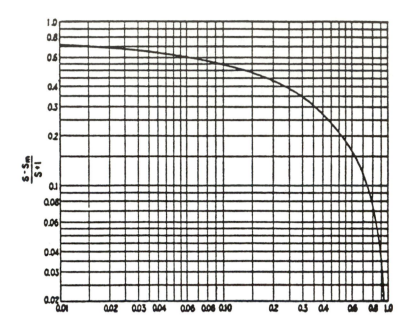

$$\frac{R - R_{min}}{R + 1}$$

FIGURE 5.30. When a system with more than two components is to be separated, we can estimate the number of stages needed using a correlation between S_{min}, the minimum number of stages needed for the separation, and R_{min}, the minimum reflux needed for separation. For further information, see eqs 5.47 through 5.49 and Table 5.6. (Reproduced from reference 45. Copyright 1940 American Chemical Society.)

Finally,

$$\ln W/W_0 = \int_{x_0}^{x} dx/(y - x) \qquad (5.53)$$

Similarly for any two components in a differential distillation,

$$A/A_0 = (B/B_0)^\alpha \qquad (5.54)$$

where A and B are the moles in the still at time t, and A_0 and B_0 are the moles at the start of the distillation. The α is the volatility of A relative to B.

EXTRACTIVE DISTILLATION—ADDING A THIRD COMPONENT TO IMPROVE VOLATILITIES

When substances are difficult to separate, we can add a third component to increase system relative volatilities. The extractive agent has an affinity for one of the components. The resulting mixture of the component and extractive agent constitutes the bottoms in the distillation column. Some typical binary systems that rely on an extractive agent are shown in Table 5.7, and a schematic of a typical extractive distillation system is shown in Figure 5.31.

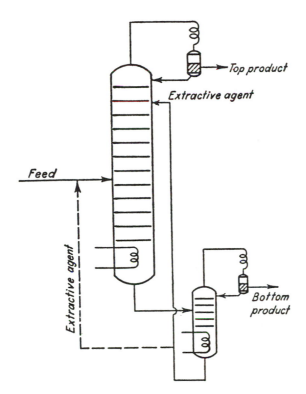

FIGURE 5.31. When a binary system is hard to separate, a third component, an extractive agent, can be added to facilitate distillation. The agent, which usually has an affinity for the bottoms product, can be added either to the feed or the top of the distillation column as shown. The agent is recycled by distillation of the bottoms in a column separate from the main column.

TABLE 5.7. Difficult-To-Separate Binary Systems and Their Extractive Agents

Binary System	Extractive Agent
Hydrochloric acid–water	Sulfuric acid
Nitric acid–water	Sulfuric acid
Ethanol–water	Glycerin
Butane–butene	Acetone or furfural
Butadiene–butene	Acetone or furfural
Isoprene–pentene	Acetone
Toluene–paraffinic hydrocarbons	Phenol
Acetone–methanol	Water

GRAPHICAL SOLUTION OF ABSORPTION IN A PLATE COLUMN

Absorption is a separation process based on a vapor–liquid system. When absorption is carried out in a column with plates or stages, the number of stages can be determined by the McCabe–Thiele method as if for a distillation column that has only a stripping section (with no rectification).

We can use Figure 5.32 for a mass balance. In absorption, we convert the mole fractions and flow rates to a solute-free basis if the system is not dilute. This process makes the mass balance lines linear. The primed quantities are created by the conversion $Y' = [Y/(1 - Y)]$ and $X' = [X/(1 - X)]$. For dilute cases, $Y = Y'$ and $X = X'$.

Then,

$$Y_{n+1}' - [(L_M'/G_M')X_n'] = Y_N' - [(L_M'/G_M')X_{n-1}'] \qquad (5.55)$$

where L_M' and G_M' are the molal mass velocities in pound-moles per hour-square foot for solute-free liquid and gas streams.

If we do a mass balance on an overall basis, then

$$Y_{n+1}' = L_M'X_n'/G_M' + Y_{N+1}' - L_M'X_N/G_M \qquad (5.56)$$

For a dilute gas

$$Y_{n+1} = L_MX_n/G_M + Y_{N+1} - L_MX_N/G_M \qquad (5.57)$$

In solving absorption graphically, we use principles similar to those for the McCabe–Thiele method. First, we locate the equilibrium line (Figure 5.33). Then we locate the operating line (its slope is L_M'/G_M' or L_M/G_M and it passes through

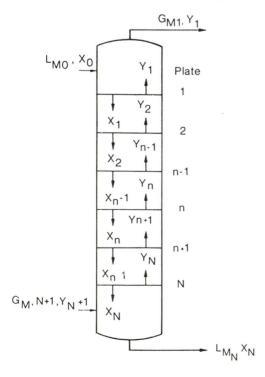

FIGURE 5.32. Absorption is a separation process, based on a liquid–vapor system, that can be treated like distillation with only a stripping process. The countercurrent process has the liquid feed, L, fed into the top of the column, and a gas, G, fed into the bottom. The McCabe–Thiele method can be used to determine the number of plates needed. (Reproduced with permission from reference 10. Copyright 1975 McGraw-Hill.)

X_0, Y, and X_n, Y_{n+1} (Figure 5.33)). Stages are then stepped off. In Figure 5.33, four stages are required to effect the separation.

In absorption, the condition of minimum liquid rate corresponds to minimum reflux. Both rates require an infinite number of stages. This value is found by drawing the operating line so that it touches the equilibrium line at X_n, Y_n (Figure 5.34).

The McCabe–Thiele approach works for plate-type absorbers because thermal effects are low (the component, found in the gas in small amounts, is being absorbed into a large mass of liquid). Example 5.5 considers the design of a plate absorber column. In this example, we use the concept of 1.5 times a minimum value of liquid flow rate as an operating parameter.

EXAMPLE 5.5. GRAPHICAL ESTIMATION OF THE NUMBER OF STAGES NEEDED FOR AN EXTRACTION

We want to remove alcohol vapor (0.01 mole fraction) from a carbon dioxide gas stream. Water for the absorption contains 0.0001 mole fraction of alcohol. A total of 227 moles of gas are to be treated per hour. The equilibrium relationship for alcohol and water is given by $Y = 1.0682 X$. For this case, how many theoretical plates would be required for 98% absorption at a liquid rate of 1.5 times minimum?

At minimum liquid rate, the operating line would intersect the equilibrium line at

$$Y = 0.01$$

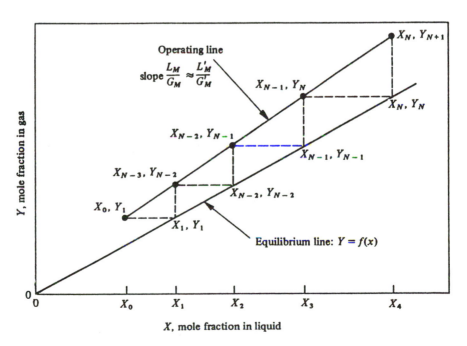

FIGURE 5.33. Application of the McCabe–Thiele method to an absorption column begins with an equilibrium line for the solute. The operating line must pass through X_0 (mole fraction of the solute in the feed of the absorbing liquid) and Y_1 (the mole fraction solute in the exiting gas), and through X_N (the mole fraction in the exiting liquid) and Y_{N+1} (the mole fraction of solute in the feed gas). (Reproduced with permission from reference 10. Copyright 1975 McGraw-Hill.)

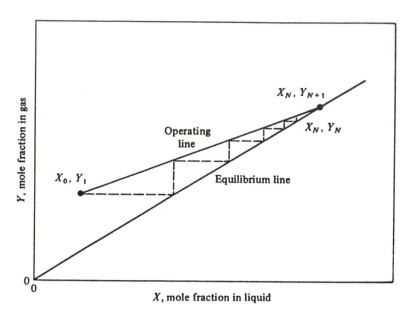

FIGURE 5.34.   In absorption, minimum liquid rate corresponds to minimum reflux in distillation. To find the number of stages needed for a minimum liquid rate, the stages must be drawn using an operating line that intersects X_n and Y_n. (Reproduced with permission from reference 10. Copyright 1975 McGraw-Hill.)

and

$$X_N = 0.01/1.0682$$

Hence, point C in Figure 5.35 is (0.009362, 0.01).

Point A is determined by X_0 and Y_1. The X_0 value is given as 0.0001 in the problem statement and Y_1 by mass balance.

Location	Moles CO_2	Moles alcohol
Bottom	224.73	2.27
Top	224.73	0.046

Then for Y_1,

$$Y_1 = 0.046/(224.73 + 0.046) = 0.000202$$

Next, the slope of AC gives the value of $L_{minimum}$,

$$L_{minimum} = (\text{moles gas})(\text{slope AC})$$
$$L_{minimum} = (224.78)(1.0588)$$

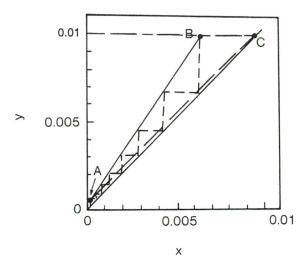

FIGURE 5.35. Given the desired liquid flow rate for an absorption process, we can determine the number of plates needed in the process by using the McCabe–Thiele method, as illustrated in Example 5.5, for which this diagram is the solution. First, point C is determined by assuming a minimum liquid flow rate. Then, point A is determined by mass balance. With the slope of line AC, the numerical value of the minimum flow rate is found, leading to the desired flow rate. Point B is determined last, and the stages are drawn. (Adapted from reference 10.)

and

$$L_{minimum} = 238 \text{ moles}$$

At 1.5 $L_{minimum}$, the L value is

$$L = (1.5)(238) = 357 \text{ moles}$$

From L we can calculate the slope of a material or mass balance line as $357/224.78$ or 1.588. Point A (0.0001, 0.000202) represents one end of the mass balance as well as the composition at the top of the tower. The other end of the line is point B (composition at the bottom of the tower). For this point, Y is 0.01. Hence, line AB passes through (0.0001, 0.000202) with a slope of 1.588 and ends at $Y = 0.01$. Now the number of stages or plates can be stepped off from either A or B to yield 9 stages.

The solution for the number of plates can also be obtained algebraically, which uses the relationship shown in Figure 5.36.

PACKED COLUMN ABSORPTION

Packed columns can also be used for absorption. Typical types of packing materials are shown in Figure 5.19; the solid packing particles efficiently disperse gas and liquid in the column. Packed-column design is more complicated than that for plate columns, involving the use of mass transfer coefficients together with an integration over the column height (Figure 5.37).

For a packed column,

$$N_A a_V = k_g a_V P(Y - Y_i) \tag{5.58}$$

and

$$N_A a_V = k_L a_V P\bar{\rho}\,(X_i - X) \tag{5.59}$$

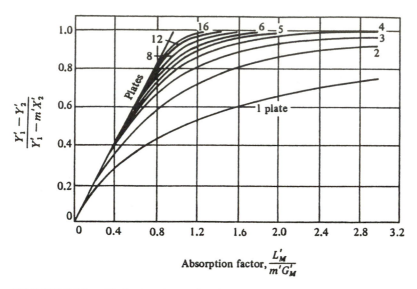

FIGURE 5.36. Algebraic solution for the number of plates in an absorption column. (Reproduced with permission from reference 10. Copyright 1975 McGraw-Hill.)

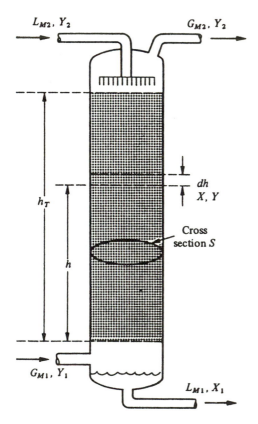

FIGURE 5.37. When a packed column is designed for absorption, the packing depth is determined by a differential approach. (Reproduced with permission from reference 10. Copyright 1975 McGraw-Hill.)

N_A is the mass transfer flux of component A, a_V is the effective mass transfer area per unit volume, k_g and k_L are the gas- and liquid-phase mass transfer coefficients (in units of moles per hour-square foot-atmosphere and feet per hour, respectively), and X_i and Y_i are the equilibrium values at the interface.

If we also use a mass balance, then

$$-d(G_M Y) = N_A a_V dh = -G_M dY - Y dG_M \tag{5.60}$$

For only one component's flux,

$$dG_M = -N_A a_V dh \tag{5.61}$$

so that

$$h_{total} = \int_1^2 dh = \int_{Y_2}^{Y_1} G_M dY / k_g P a_V (1 - Y)(Y - Y_i) \tag{5.62}$$

or

$$h_{total} = \int_1^2 dh = \int_{x_2}^{x_1} L_M dX / k_L a_V (1 - X)(X_i - X)\bar{\rho} \tag{5.63}$$

If we use the concept of a Y_{BM},

$$Y_{BM} = [(1 - Y) - (Y - Y_i)]/\ln [(1 - Y)/(Y - Y_i)] \tag{5.64}$$

then,

$$H_G = G_M / k_g a_V P Y_{BM} \tag{5.65}$$

and

$$N_G = \int_{Y_2}^{Y_1} Y_{BM} dY / (1 - Y)(Y - Y_i) \tag{5.66}$$

so that

$$h_{total} = (H_G)_{average} (N_G) = (H_G)_{average} \int_{Y_2}^{Y_1} Y_{BM} dY / (1 - Y)(Y - Y_i) \tag{5.67}$$

We want to use overall mass transfer coefficients so that the interfacial compositions are not needed. To do this, we use values of liquid X^* or vapor Y^* in equilibrium with vapor (for X^*) and liquid.

Then,

$$N_A = K_{OG} P(Y - Y^*) \tag{5.68}$$

and

$$N_A = K_{OL}(X^* - X) \tag{5.69}$$

The values of K_{OG} and K_{OL} are related to k_g and k_L by

$$1/K_{OG} = 1/k_g + m(P/\bar{\rho})(1/k_L) \tag{5.70}$$

and

$$1/K_{OL} = 1/k_L + (\bar{\rho}/m'P)(1/k_g) \tag{5.71}$$

where m is the equilibrium-line slope from X, Y^* to X_i, Y_i, and m' is the equilibrium-line slope from X^*, Y to X_i, Y_i.

Basically,

$$L_m(X_1 - X_2) = G_m(Y_1 - Y_2) = K_{OG}a_VPh_{total}(Y - Y^*)_{lm} \tag{5.72}$$

and

$$(Y - Y^*)_{lm} = [(Y - Y^*)_1 - (Y - Y^*)_2]/\ln (Y - Y^*)_1/(Y - Y^*)_2 \tag{5.73}$$

where the subscript lm refers to the logarithmic mean value.

The same is true for the liquid side:

$$G_m(Y_1 - Y_2) = L_m(X_1 - X_2) = \bar{\rho} \, K_{OL}a_vh_{total}(X^* - X)_{lm} \tag{5.74}$$

and

$$(X^* - X)_{lm} = [(X^* - X)_1 - (X^* - X)_2]/\ln (X^* - X)_1/(X^* - X)_2 \tag{5.75}$$

Another method is to use the concept of a dilute gas so that

$$H_{OG} = H_G + (mG_m/L_m)H_L \tag{5.76}$$

and

$$H_{OL} = H_L + (L_m/m'G_m)H_G \tag{5.77}$$

Likewise the N_{OG} and N_{OL} values become

$$N_{OG} = [1/(1 - K)] \ln [(1 - K)(Y_1 - mX_2)/(Y_2 - mX_2) + K] \tag{5.78}$$

and

$$N_{OL} = [1/(1 - K')] \ln [(1 - K')(X_1 - Y_2/m)/(X_2 - Y_2/m) + K'] \tag{5.79}$$

where

$$K = mG_m/L_m$$

and

$$K' = L_m/m'G_m$$

The expressions for N and K_{OG} or N_{OL} and K' have been correlated by plots such as Figure 5.38.

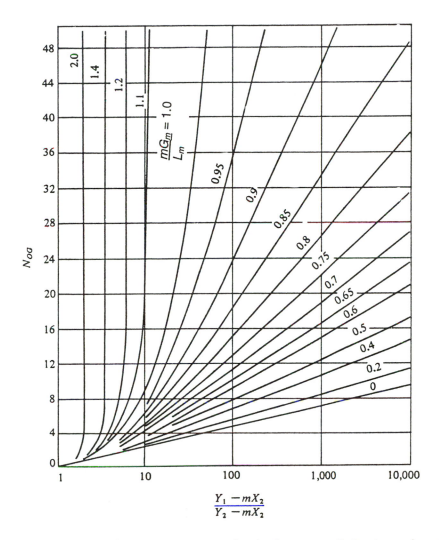

FIGURE 5.38. Given a constant ratio of molar flow rates, mG_m/L_m, the number of transfer units needed for absorption can be found from this correlation. (Reproduced with permission from reference 10. Copyright 1975 McGraw-Hill.)

EXAMPLE 5.6. ESTIMATING THE PACKED COLUMN HEIGHT NEEDED FOR ABSORPTION

We want to remove alcohol vapor from carbon dioxide using a packed column containing 2.54-cm Raschig rings at 40 °C. The vapor (0.01 mole fraction alcohol) needs to have 98% recovery of alcohol. The water used for absorption has 0.0001 mole fraction alcohol. The G_m at the tower bottom is 0.0422 kg-mol/s-m^2. The ratio of L_m to G_m is 1.5264. Equilibrium is given by $Y = 1.0682x$. Values of H_G and H_L are 0.54 and 0.30 m, respectively.

We can calculate the column height using three methods: (1) assuming Y_{BM} (eqs 5.60–5.75), (2) assuming dilute gas conditions (eqs 5.76 and 5.77), and (3) using the correlation in Figure 5.38.

CASE I. DETAILED SOLUTION. First, we calculate mG_m/L_m.

$$mG_m/L_m = 1.0682/1.5264 = 0.70$$

Next, we calculate H_{OG} using eq 5.76.

$$H_{OG} = H_G + (mG_m/L_m)H_L = 0.54 \text{ m} + (0.7)(0.30 \text{ m})$$

$$H_{OG} = 0.75 \text{ m}$$

Now, we obtain X and Y values at the top and bottom of the column.

$$Y_{bottom} = 0.01 \text{ (given)}$$

$$Y_{top} = 0.02(0.01/(1 - 0.01)) = 0.000202$$

and

$$X_{top} = 0.0001$$

The X at the bottom can be obtained by the mass balance. The alcohol in the gas stream fed to the absorber is

$$(0.000202/0.02)(0.042) \text{ kg-mol/s-m}^2 = 0.000422 \text{ kg-mol/s-m}^2$$

$$\text{alcohol absorbed} = 0.000424(0.98) \text{ kg-mol/s-m}^2$$

$$\text{alcohol absorbed} = 0.000414 \text{ kg-mol/s-m}^2$$

Using the alcohol absorbed, we can calculate X at the bottom as

$$X_{bottom} = [0.000414 + (0.0001)(1.5264)(0.042)]\text{kg-mol/s-m}^2/$$
$$[(1.5264)(0.042) + 0.000414] \text{ kg-mol/s-m}^2$$

$$X_{bottom} = 0.006542$$

The Y^* values corresponding to the X values are

$$X_{bottom} = 0.006542$$

$$Y^*_{bottom} = (1.0682)(0.006542) = 0.006988$$

$$X_{top} = 0.0001 \text{ (given)}$$

$$Y^*_{top} = (1.0682)(0.0001) = 0.0010682$$

With these values and eq 5.73

$$(Y - Y^*)_{lm} = [(Y - Y^*)_{bottom} - (Y - Y^*)_{top}]/\ln(Y - Y^*)_{bottom}/(Y - Y^*)_{top}$$
$$(Y - Y^*)_{lm} = [(0.01 - 0.006988) - (0.000202 - 0.0010682)]/$$
$$\ln[(0.01 - 0.006988)/(0.000202 - 0.00010682)]$$
$$(Y - Y^*)_{lm} = 0.0008443$$

Then, from eq 5.72,

$$(G_mY_1 - G_mY_2)/K_{OG}a_VP(Y - Y^*)_{lm} = h_{total}$$

However, we know (from eq 5.67 and 5.65) that

$$K_{OG}a_VP = (G_M)_{average}/H_{OG}Y^*_{BM}$$

so that

$$K_{OG}a_VP = \frac{0.042}{(0.75)(0.9956)} = 0.0562 \text{ kg-mol/s-m}^3$$

$$h_t = (0.0422)(0.01)/0.0562(0.008443) = 8.74 \text{ m}$$

CASE 2. USING AN ANALYTICAL TECHNIQUE.

$$(Y_1 - mX_2)/(Y_2 - mX_2) = [0.01 - (1.0682)(0.0001)]/$$
$$[(0.000202) - (1.0682)(0.0001)]$$

$$(Y_1 - mX_2)/(Y_2 - mX_2) = 103.94$$

Then, from eq 5.78,

$$N_{OG} = [1/(1 - K)] \ln [(1 - K)(Y_1 - mX_2)/(Y_2 - mX_2) + K]$$
$$N_{OG} = \ln [(1 - 0.7)(103.94) + 0.7]/(1 - 0.7)$$
$$N_{OG} = 11.57$$
$$h_t = (N_{OG})(H_{OG}) = (11.57)(0.75 \text{ m}) = 8.68 \text{ m}$$

CASE 3. USING FIGURE 5.38. We can read $(Y_1 - mX_2)/(Y_2 - mX_2)$ as the abscissa of 103.94. The ordinate is 11.5. Therefore

$$h_t = (11.5)(0.75 \text{ m}) = 8.69 \text{ m}$$

In this example, the less precise methods give column heights that compare favorably to that found by the rigorous method of case 1 (8.68 and 8.69 m compared to 8.74 m).

Absorption in multicomponent systems and nonisothermal systems can be handled by complicated techniques given in other sources (*10*).

EXTRACTION

In extraction processes the solution to be separated is put in contact in stages with an extracting solvent. The solvent-rich phase leaving the stage is the extract, and the solvent-lean phase is the raffinate. There are many analogies between extraction and distillation (Table 5.8).

Graphical solutions for extraction can be carried out on (a) ternary diagrams in triangular coordinates, (b) distribution diagrams with rectangular coordinates of

TABLE 5.8.   Analogies Between Distillation and Extraction

Extraction	Distillation
Addition of solvent	Addition of heat
Solvent mixer	Reboiler
Removal of solvent	Removal of heat
Solvent separator	Condenser
Solvent-rich solution saturated with solvent	Vapor at the boiling point
Solvent-rich solution containing more solvent than that required to saturate it	Superheated vapor
Solvent-lean solution containing less solvent than that required to saturate it	Liquid below the boiling point
Solvent-lean solution saturated with solvent	Liquid at the boiling point
Two-phase liquid mixture	Mixture of liquid and vapor
Selectivity	Relative volatility
Change of temperature	Change of pressure

weight fraction of component C (the solute) in component B versus weight fraction of component C in component A, (X_{CB} vs. X_{CA}), and (c) a Janecke diagram in which $X_B/(X_A + X_C)$ is plotted against $X_C/(X_A + X_C)$ where X_A, X_B, X_C are weight fractions of A, B, and C, respectively.

Of the three methods, let us look at a graphical solution using a ternary diagram for a countercurrent, multiple-contact extraction system. (Descriptions of Janecke and distribution diagrams are given in references 5 and 14.) The extraction system, which we shall examine, has both an extract and raffinate reflux (Figure 5.39).

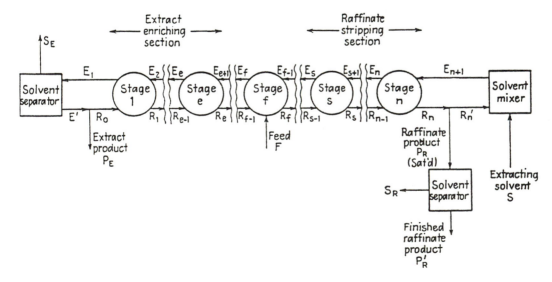

FIGURE 5.39.   A countercurrent liquid–liquid extraction process can be drawn for mass balance purposes. This system has both an extract and raffinate reflux.

We shall look at the extract enriching section first. By mass balance only around the solvent separator

$$E_1 = S_E + E' \tag{5.80}$$

and

$$E_1 = S_E + P_E + R_o \tag{5.81}$$

where E_1 is the extract, R is the raffinate, P is the extract product, and S is the solvent. In the extract section, S_E and P_E are the same regardless of where we place boundaries for the mass balance. We therefore define a quantity, Q, as

$$Q = S_E + P_E \tag{5.82}$$
$$E_1 = Q + R_o \tag{5.83}$$

A material balance about the extract side of the system for any stage gives

$$E_{e+1} = S_E + P_E + R_e \tag{5.84}$$

and

$$S_E + P_E = Q = E_{e+1} - R_e \tag{5.85}$$

To start the graphical solution of the number of stages needed for the extraction, we locate S_E and P_E on a triangular diagram (Figure 5.40), connect them with a line, and find Q on line $S_E P_E$. Next we determine the first stage of the stepwise solution by going from E_1 to R_1 on the dotted tie line in the phase diagram equilibrium and then to E_2 by connecting R_1 with Q (material balance and line). We repeat the procedure until the vicinity of F, the feed point, is reached.

Similarly, we can start the calculation of the number of stages needed for the raffinate end by doing a material balance around the solvent mixer,

$$S - P_R = W \tag{5.86}$$

The material balance around the entire raffinate end also gives

$$S - P_R = W = E_{s+1} - R_s \tag{5.87}$$

To find the number of stages graphically, we use the triangular diagram once again (Figure 5.41). First we locate the W point by using eq 5.87. Then, the solution proceeds in a stepwise manner: R_n is used to locate E_n by tie line (equilibrium); then, a line drawn through E_n and W locates R_{n-1} (material balance operating line). We repeat the procedure until the vicinity of the feed point is reached.

The feed stage is located because $W + F = Q$ (that is, F, Q, and W all lie on a straight line). This fact is used to reconcile the extract end (Q) with the raffinate end (W) as follows: Point Q is used as an operating point (Figure 5.40) until a tie line crosses FQ (line connecting points F and Q). When this situation occurs, W is used as an operating point until the raffinate end is completed.

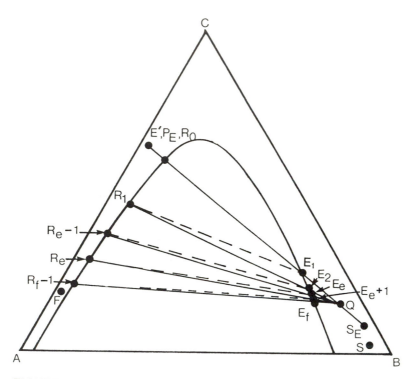

FIGURE 5.40.   Stage calculation for the extraction end of an extraction process. Calculations for an extraction system (schematic shown in Figure 5.38) may be done using a ternary diagram (each point represents 100% of the component), with the number of stages determined in a stepwise fashion. In the ternary diagram shown here the solute C is soluble in both A and B, but A and B are relatively insoluble in each other.

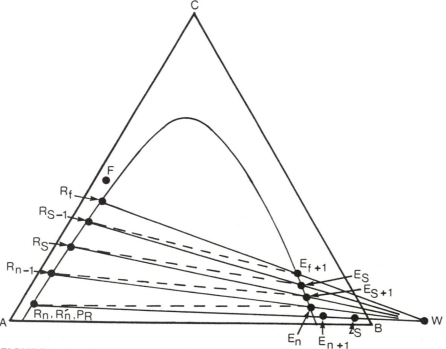

FIGURE 5.41.   Stage calculation for the raffinate end of an extraction process.

As with distillation, minimum reflux (infinite number of stages) and minimum number of stages (total reflux) occur in extraction. Minimum reflux takes place in the extract section when an extended tie line meets Q as close to S_E as possible (Figure 5.40). For the raffinate section, an extended line will also meet the W point as close to S as possible (Figure 5.41). For total reflux, the points S, S_E, Q, and W must all coincide.

IDEAL AND REAL STAGES AND STAGE INEFFICIENCIES

The principles we applied in determining the number of stages needed for distillation, absorption, and extraction apply to other equilibrium stage operations, such as leaching, crystallization, and adsorption. We have calculated the number of theoretical stages required, yet no staged separation unit is 100% efficient. Therefore, we must make allowances for process inefficiencies and increase the number of actual stages required. Typical efficiencies are usually about 60–70%. We can divide the number of theoretical stages by the efficiency to determine the actual number of stages needed. The correlations for estimating efficiencies are empirical and apply only to limited systems (see references 5, 11, and 15).

MECHANICAL SEPARATION PROCESSES

Another class of mass transfer involves mechanical separation. Included in this class are filtration, centrifugation, cyclone separation, sedimentation, and electrostatic precipitation (for examples of equipment, *see* Figures 5.42–5.52). All of these machines have driving forces or separating agents that are mechanically based (Table 5.9).

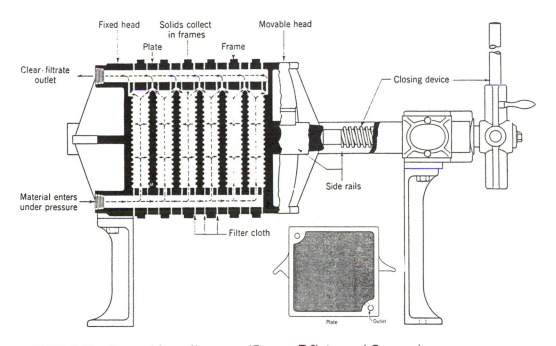

FIGURE 5.42.   Plate-and-frame filter press. (Courtesy T. Shriver and Company.)

TABLE 5.9. Mechanical Separations

Process	Phases	Separation Method
Filtration	Liquid–solid	Pressure reduction
Centrifugation	Liquid–solid or liquid–liquid	Centrifugal force
Sedimentation	Liquid–solid	Gravity
Cyclone separator	Gas–solid or gas–liquid	Flow
Electrostatic precipitator	Gas–solid	Electric field
Demister	Gas–solid or gas–liquid	Pressure reduction

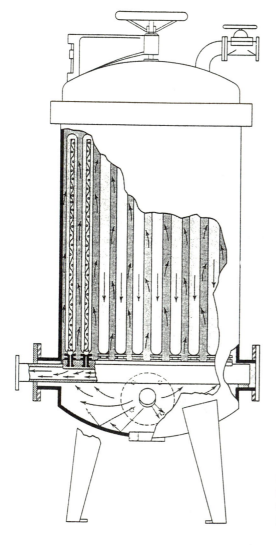

FIGURE 5.43. Vertical leaf filter. (Courtesy Industrial Filter and Pump Manufacturing Company.)

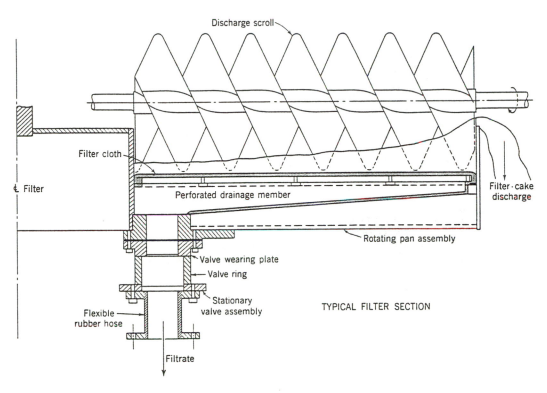

FIGURE 5.44. Rotary horizontal vacuum filter. (Courtesy Filtration Engineers Division, American Machine and Metals, Inc.)

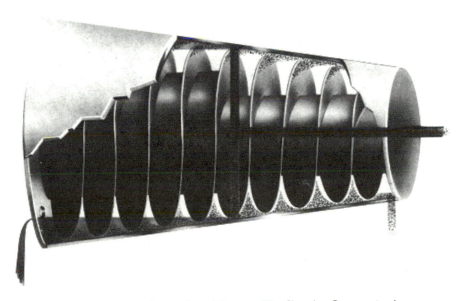

FIGURE 5.45. Solid-bowl centrifuge. (Courtesy The Sharples Corporation.)

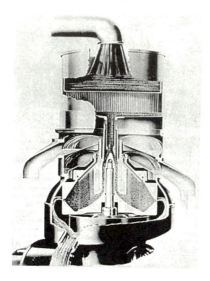

FIGURE 5.46. Disk-bowl centrifuge. (Courtesy DeLaval Separator Company.)

TABLE 5.10. Comparison of Heat and Mass Transfer Operations

Heat Transfer	Mass Transfer
Steady-state heat transfer	Steady-state molecular diffusion
Unsteady-state heat transfer	Unsteady-state molecular diffusion
Convective heat transfer (heat transfer coefficient)	Convective mass transfer (mass transfer coefficients)
Convective heat transfer (heat transfer coefficient)	Equilibrium staged operations (convective mass transfer using departure from equilibrium as a driving force)
Radiative heat transfer (not analogous with other transfer processes)	Mechanical separations (not analogous with other transfer processes)

The mass transfer of mechanical separation has no counterpart in heat transfer. However, this fact is also true for radiation heat transfer, which has no mass transfer analog (Table 5.10).

FILTRATION

Filtration is one of the most widespread separation processes. It involves the flow of a slurry or solution containing particulate matter through a porous medium. The basic equation for filtration (*16,17*) is

$$-(\Delta P)g_c/L = 180[(1 - \epsilon)^2/\epsilon^3][\mu V_s/D_p^2] \tag{5.88}$$

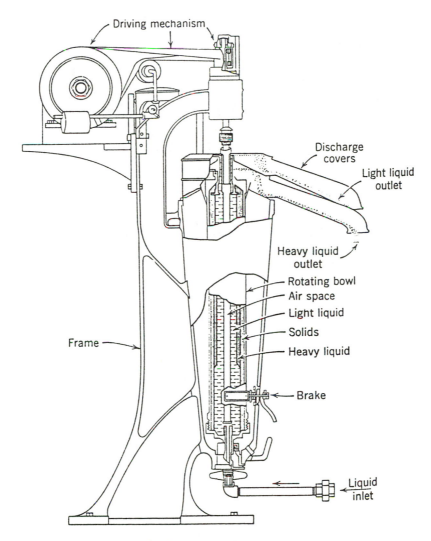

FIGURE 5.47. Tubular-bowl centrifuge. (Reproduced with permission from reference 52. Copyright 1960 Wiley.)

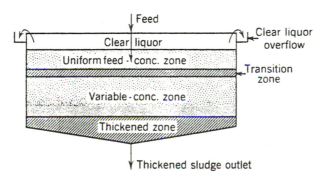

FIGURE 5.48. Sedimentation is a diffusion of particles through several zones. Movement takes place by the action of gravity on the solid particles. (Reproduced with permission from reference 52. Copyright 1960 Wiley.)

FIGURE 5.49. A continuous thickener separates solids from liquid. (Courtesy Dorr-Oliver Inc.)

where L is the thickness of the filter, V_s is the superficial velocity, ϵ is porosity or fraction of total volume that is void, and D_p is particle diameter. We can define V_s further as

$$V_s = 1/A[d\overline{V}/d\theta] = -(\Delta P)g_c\epsilon^3/5L\mu S_o^2(1 - \epsilon)^2 \qquad (5.89)$$

where $d\overline{V}/d\theta$ is the volumetric filtration rate, S_o is the specific area of particle in square feet per cubic feet of volume, and A is the cross-sectional area of the filter.

FIGURE 5.50. A plate-type electrostatic precipitator (ionizer) charges solid particles. (Reproduced with permission from reference 49. Copyright 1968 Pergamon.)

FIGURE 5.51. A plate-type electrostatic precipitator (collector) takes up particles charged by the ionizer shown in Figure 5.50. (Reproduced with permission from reference 49. Copyright 1968 Pergamon.)

CENTRIFUGATION

Another widely used mechanical separation is centrifugation. Application of centrifugal force brings about a separation of a liquid–solid or liquid–liquid system. The basic force balance for the centrifuge is

$$dV/d\theta = rw^2[(\rho_s - \rho)/\rho_s] - C_D\rho V^2 S/2m \qquad (5.90)$$

where $dV/d\theta$ is acceleration, θ is time, V is velocity, w is angular velocity, ρ_s is solid particle density, ρ is fluid density, C_D is the drag coefficient of the particle, S is the projected area perpendicular to flow, and m is particle mass. From this force balance, the equation governing centrifugation is

$$\ln r_2/r_1 = [rw^2(\rho_s - \rho)D_p^2/18\mu]\theta \qquad (5.91)$$

or

$$\ln r_2/r_1 = [rw^2(\rho_s - \rho)D_p^2\theta/18\mu]\overline{V}/Q \qquad (5.92)$$

where θ is time, $\overline{V}$ is volume, and Q is volumetric flow rate.

For any two centrifuges that perform the same function (*18*),

$$Q_1/\Sigma_1 = Q_2/\Sigma_2 \qquad (5.93)$$

where Σ is the surface area of a sedimentation device that would perform the same separation as the centrifuge. Various centrifuges are shown in Figures 5.45–5.47.

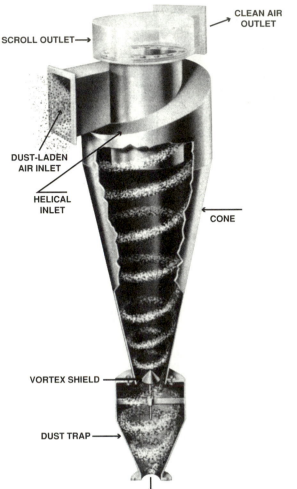

SCROLL OUTLET→

**CLEAN AIR
OUTLET**

**DUST-LADEN
AIR INLET**

**HELICAL
INLET**

CONE

VORTEX SHIELD

DUST TRAP →

FIGURE 5.52. The cyclone separator removes solid particles from a flowing stream. (Courtesy The Ducon Company.)

Other mechanical separation devices are shown in Figures 5.48–5.52. All mechanical separation processes have given efficiencies for their removal abilities. A typical efficiency curve is shown for a cyclone separator in Figure 5.53. The curve shows a 50% efficiency when we select D_p equal to D_{cut} (i.e., 50% of a chosen particle size is to be removed). The curve shows that much larger particles, two, four, six, and even eight times larger, can still pass through the system. This behavior applies to all of the mechanical separation processes, albeit with different efficiency curves.

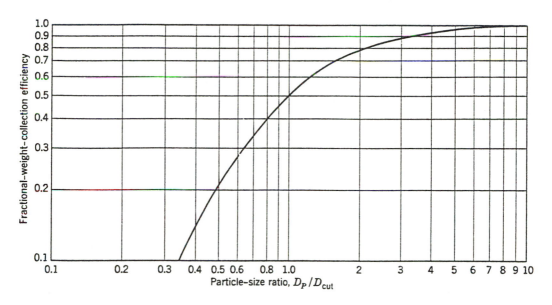

FIGURE 5.53. Efficiency curve for a cyclone separator. (Reproduced with permission from reference 52. Copyright 1960 Wiley.)

REFERENCES

1. Onsager, L. *Phys. Rev.* **1931**, *37*, 405.
2. Onsager, L. *Phys. Rev.* **1931**, *38*, 2265.
3. Carslaw, H. S.; Jaeger, J. C. *Heat Conduction in Solids;* Oxford University: Oxford, England, 1959, out of print.
4. Crank, J. *The Mathematics of Diffusion;* Oxford University: Oxford, England, 1956.
5. Chilton, T. H.; Colburn, A. P. *Ind. Eng. Chem.* **1934**, 26, 1183.
6. Sherwood, T. K.; Pigford, R. L. *Absorption and Extraction.*, 2nd ed.; McGraw-Hill: New York, 1954.
7. Gilliland E. R.; Sherwood, T. K. *Ind. Eng. Chem.* **1934**, *26*, 516.
8. Frossling, N. *Gerlands Beitr. Geophys.* **1938**, *52*, 170.
9. Brian, P. L. T.; Hales, H. B. *AIChE J.* **1969**, *15*, 419.
10. Sherwood, T. K.; Pigford, R. L.; Wilke, C. R. *Mass Transfer;* McGraw Hill: New York, 1975.
11. Johnson, A. I.; Besic, F.; Hamielec, A. E. *Can. J. Chem.* **1969**, *47*, 559.
12. Levich, V. G. *Physicochemical Hydrodynamics;* Prentice-Hall: Englewood Cliffs, NJ, 1962.
13. Cornet, I.; Kaloo, U. *Tr. Mezhdunar, Kongr. Korroz. Met., 3rd.* **1966**, *3*, 83.
14. Underwood, A. J. V. *J. Inst. Pet.* **1946**, *32*, 614.
15. Underwood, A. J. V. *Trans. Inst. Chem. Eng.* **1932**, *10*, 112.
16. Carman, P. C. *J. Soc. Chem. Ind., London* **1938**, *57*, 225T.
17. Kozeny, J. S. *Akad. Wiss. Wien Math-naturw.* **1927**, *136*, 271.
18. Ambler, C. M. *Chem. Eng. Prog.* **1952**, *48*, 150.
19. Rayleigh, L. *Philos. Mag.* **1902**, *4*, 521.
20. Sieder, E. N.; Tate, G. E. *Ind. Eng. Chem.* **1936**, *28*, 1429–1435.

21. Hikita, H.; Nakanishi, K.; Asai, S. *Kagaku Kogaku* **1959**, *23*, 28.
22. Nusselt, W. Z. *VDI-Z* **1923**, *67*, 206.
23. Kramers, H.; Kreyger, P. J. *Chem. Eng. Sci.* **1956**, *6*, 42.
24. Harriott, P. *AIChE J.* **1962**, *8*, 93.
25. McAdams, W. H.; Drew, T. B.; Bays, G. S., Jr. *Trans. ASME* **1940**, *62*, 627.
26. Vogtlander, P. H.; Bakker, C. A. P. *Chem. Eng. Sci.* **1963**, *18*, 583.
27. Johnson, T. R.; Joubert, P. N. *J. Heat Transfer* **1969**, *February*, 91.
28. Kestin, J.; Wood, R. T. *Am. Soc. Mech. Eng. [Pap.]* 70-WA/HT-3, 1970.
29. Sogin, H. H.; Subramanian, V. S. *Am. Soc. Mech. Eng. [Pap.]* 60-WA-193, 1960.
30. Bar-Ilan, M.; Resnick, W. *Ind. Eng. Chem.* **1957**, *49*, 313.
31. Gamson, B. W.; Thodos, G.; Hougen, O. A. *Trans. Am. Inst. Chem. Eng.* **1943**, *39*, 1.
32. De Acetis, J.; Thodos, G. *Ind. Eng. Chem.* **1960**, *52*, 1003.
33. Hobson, M.; Thodos, G. *Chem. Eng. Prog.* **1951**, *47*, 370.
34. McConnachie, J. T. L.; Thodos, G. *AIChE J.* **1963**, *9*, 60.
35. Hobson, M.; Thodos, G. *Chem. Eng. Prog.* **1949**, *45*, 517.
36. Williamson, J. E.; Bazaire, K. E.; Geankoplis, C. J. *Ind. Eng. Chem. Fundam.* **1963**, *2*, 126.
37. Gaffney, B. J.; Drew, T. B. *Ind. Eng. Chem.* **1950**, *42*, 1126.
38. Evans, G. C.; Gerald, C. F. *Chem. Eng. Prog.* **1953**, *49*, 135.
39. Dropkin, D.; Carmi, A. *Trans. ASME* **1957**, *79*, 474.
40. Eisenberg, M.; Tobias, C. W.; Wilke, C. R. *Chem. Eng. Prog. Symp. Ser.* **1955**, *51*(16), 1.
41. Kays, W. M.; Bjorkland, I. S. Department of Mechanical Engineering, Stanford University, Technical Report 27, 1955.
42. Kays, W. M.; Bjorkland, I. S. *Am. Soc. Mech. Eng. [Pap.]* 56-A-71, 1956.
43. Seban, R. A.; Johnson, H. A. *NASA Memo.* 4-22-59W, 1958.
44. Theodorsen, T.; Regier, A. *Natl. Advis. Comm. Aeronaut., Rep.* 793, 1945.
45. Gilliland, E. R. *Ind. Eng. Chem.* **1940**, *32*, 1220.
46. McCabe, W. L.; Smith, J. C. *Unit Operations of Chemical Engineering*, 2nd ed.; McGraw-Hill: New York, 1967.
47. Othmer, D. F.; White, R. E.; Trueger, E. *Ind. Eng. Chem.* **1941**, *33*, 1240.
48. McCabe, W. L.; Smith, J. C.; Harriott, P. *Unit Operations of Chemical Engineering*, 4th ed.; McGraw-Hill: New York, 1985.
49. Coulson, J. M.; Richardson, J. F. *Chemical Engineering;* Pergamon: London, 1968; Vols. 1 and 2.
50. Geankoplis, C. J. *Transport Processes and Unit Operations*, 2nd ed.; Allyn and Bacon: Boston, MA, 1989.
51. Treybal, R. E. *Mass Transfer Operations*, 3rd ed.; McGraw-Hill: New York, 1979.
52. Foust, A. S.; Wenzel, L. A.; Clump, C. W.; Maus, L.; Anderson, L. *Principles of Unit Operations*, 2nd ed.; reprint of 1980 ed.; Krieger, 1990.

FURTHER READING

Bennett, C. O.; Myers, J. E. *Momentum, Heat, and Mass Transfer*, 3rd ed.; McGraw-Hill: New York, 1982.
Bird, R. B.; Stewart, W. E.; Lightfoot, E. N. *Transport Phenomena;* John Wiley and Sons: New York, 1960.
Fahien, R. *Fundamentals of Transport Phenomena;* McGraw-Hill: New York, 1983.
Greenkorn, R. A.; Kessler, D. P. *Transfer Operations;* McGraw-Hill: New York, 1972, out of print.
King, C. J. *Separation Processes*, 2nd ed.; McGraw-Hill: New York, 1979.

Rohsenow, W. M.; Choi, H. *Heat, Mass, and Momentum Transfer;* Prentice-Hall: Engle-
wood Cliffs, NJ, 1961.
Sherwood, T. K.; Pigford, R. L. *Absorption and Extraction;* McGraw-Hill: New York, 1954,
out of print.
Sherwood, T. K.; Pigford, R. L.; Wilke, C. R. *Mass Transfer,* 2nd ed.; McGraw-Hill: New
York, 1989.

6

Applied Reaction Kinetics and Reactor Design

Chemical kinetics is familiar to chemists from their curriculum. A large gap sometimes exists between this knowledge and chemical kinetics as applied in the process industries. Our goal in this chapter is to use the chemist's knowledge of chemical kinetics to develop an understanding of applied kinetics. We shall concentrate on nonisothermal reactions, heterogeneous systems, elements of reactor design, and methods and techniques needed to understand the operation of industrial reactors.

A REVIEW OF CHEMICAL KINETICS AND CHEMICAL EQUILIBRIUM

Chemical reactions take place because of chemical affinity. The reaction rate r in general is

$$r = -\frac{1}{V}\frac{dn_A}{dt} \tag{6.1}$$

where n_A is moles of reactant A, V is volume, and t is time. If the volume is constant

$$r = -\frac{dC_A}{dt} \tag{6.2}$$

where C_A is the concentration of reactant A in moles per volume.

Chemical reactions proceed at a rate r, reaching a condition of equilibrium at which the rate approaches zero asymptotically on a plot of rate versus time (Figure 6.1). On a plot of conversion versus time, equilibrium occurs at the final percentage conversion attained. It indicates the likelihood that a reaction will produce an appreciable amount of product (Figure 6.2).

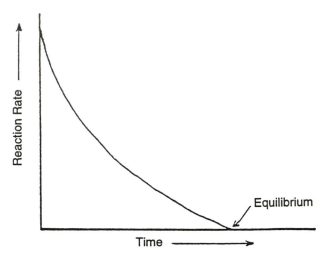

FIGURE 6.1. Equilibrium occurs as the rate of a reaction approaches zero, as shown in a plot of reaction rate vs. time. (Adapted from reference 2.)

Hence, chemical equilibrium calculations are important when evaluating chemical reactions. A chemical equilibrium constant K is defined by the relationship

$$\Delta F° = -RT \ln K \tag{6.3}$$

where $\Delta F°$ is the change in Gibbs free energy between products and reactants at a standard state. K can be defined by its relation to activities. If a reaction can be written as

$$bB + cC \rightleftarrows dD + eE \tag{6.4}$$

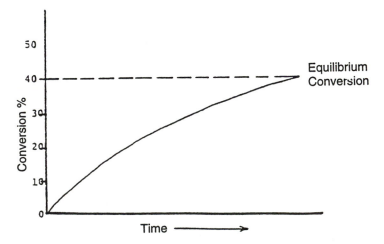

FIGURE 6.2. As a reaction approaches equilibrium, the conversion rate reaches the equilibrium value.

then the equilibrium constant K in terms of activities is

$$K = \frac{(a_D)^d (a_E)^e}{(a_B)^b (a_C)^c}$$ (6.5)

where activity is represented by a.

RELATIONSHIP OF ACTIVITY TO FUGACITY

As introduced in Chapter 2, activity a is related to another term fugacity f by

$$a = \frac{f}{f^0}$$ (6.6)

where the superscript refers to a standard state. Fugacity is defined as

$$\frac{d(\ln f)}{dP} = \frac{\bar{v}}{RT}$$ (6.7)

where $\bar{v}$ is volume per mole, R is the gas constant, T is the absolute temperature, and P is the pressure. Fugacity is also related to chemical potential by the relationship

$$\mu = RT \ln f + \theta$$ (6.8)

where θ is a temperature function.

The value of the chemical equilibrium constant K, therefore, is fixed by the parameter used. For the chemical reaction of eq 6.4,

$$K_g = \frac{(g_D)^d (g_E)^e}{(g_B)^b (g_C)^c}$$ (6.9)

where g can be activity a, fugacity f, partial pressure P, mole fraction y, or concentration c.

The following problem offers a review of chemical equilibrium principles and the use of K.

EXAMPLE 6.1. CALCULATING EQUILIBRIUM CONCENTRATIONS, GIVEN K

For the water–gas reaction,

$$CO\ (gas) + H_2O\ (gas) \rightleftarrows CO_2\ (gas) + H_2\ (gas)$$

the equilibrium constant K_y is 1.0. We must calculate the equilibrium fraction of steam decomposed to carbon dioxide and hydrogen if two moles of nitrogen are included in the reactor and one mole each of carbon monoxide and water are present initially.

Assuming that the number of moles of H_2 at equilibrium equal z, we can describe the moles of the reactants and products remaining at equilibrium as shown in the table below:

Component	Moles	Mole fraction
H_2	z	$z/4$
CO_2	z	$z/4$
CO	$1-z$	$(1-z)/4$
H_2O	$1-z$	$(1-z)/4$

$$\text{reactants} + \text{products} + \text{nitrogen} = z + z + (1-z) + (1-z) + 2 = 4$$

Then,

$$K_y = \frac{(\text{mole fraction } CO_2)(\text{mole fraction } H_2)}{(\text{mole fraction } CO)(\text{mole fraction } H_2O)}$$

$$1.0 = \frac{\left(\dfrac{z}{4}\right)\left(\dfrac{z}{4}\right)}{\left(\dfrac{1-z}{4}\right)\left(\dfrac{1-z}{4}\right)}$$

$$z = 0.50$$

REACTION ORDER AND EFFECT OF TEMPERATURE ON REACTION RATE

During the rate-dependent portion of the chemical reaction, the rate is related to the concentrations of the reactants raised to a power (reaction order). For a simple reaction

$$mA + pB + qC + \ldots zZ \tag{6.10}$$

the rate is

$$r = -\frac{dC_A}{dt} = k_r(C_A)^m(C_B)^p(C_C)^q \cdots (C_Z)^z \tag{6.11}$$

The Cs represent the concentrations of the reactants, and the powers are the appropriate reaction orders. The k_r is the specific reaction rate coefficient that resembles the transport coefficients (viscosity, thermal conductivity) for physical processes. The k_r is an exponential function of temperature as shown by experimental data (Figure 6.3). Over moderate temperature ranges, k_r can be found by using the Arrhenius equation:

$$k_r = Ae^{-E/RT} \tag{6.12}$$

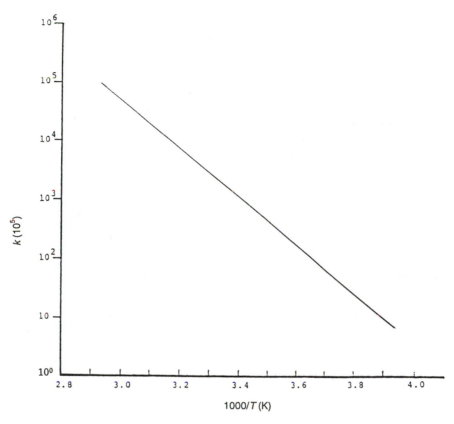

FIGURE 6.3. The specific rate of reaction k coefficient is an exponential function of temperature. (Adapted from reference 2.)

where E is the activation energy in kilojoules per mole, R is the gas constant, T is the temperature in K, and A is the frequency or preexponential factor. Over larger temperature ranges, the Arrhenius equation takes the form

$$k_r = T^m e^{\Delta S/R} e^{-\Delta H/RT} \tag{6.13}$$

where m is some integral multiple of 0.5, and ΔS and ΔH are the entropy and enthalpy changes associated with the reaction. In many instances, the value of m is unity.

CHEMICAL REACTION KINETICS CATEGORIES

We have reviewed chemical transformations as functions of time and temperature. Other factors, however, must be taken into account to categorize chemical reaction kinetics. One factor is whether the system is homogeneous (takes place in one phase such as a gas or liquid) or heterogeneous (takes place in more than one phase such as between a gas and solid, gas and liquid, liquid and liquid, etc.). Another factor is whether the reaction is enhanced with a catalyst (a substance that influences the rate of reaction but is not altered). If we consider other factors besides time and temperature, then we can sort chemical reaction kinetics into the following categories:

(1) homogeneous, isothermal, uncatalyzed;
(2) homogeneous, isothermal, catalyzed;
(3) homogeneous, nonisothermal, uncatalyzed;
(4) homogeneous, nonisothermal, catalyzed;
(5) heterogeneous, isothermal, uncatalyzed;
(6) heterogeneous, isothermal, catalyzed;
(7) heterogeneous, nonisothermal, uncatalyzed; and
(8) heterogeneous, nonisothermal, catalyzed.

The chemistry curriculum generally covers category 1 thoroughly and categories 2 and 6 reasonably well. The remaining categories receive little or no attention and are explored in this chapter.

HOMOGENEOUS, ISOTHERMAL, UNCATALYZED REACTION KINETICS

We will first determine the order and specific reaction rates for isothermal reactions. The order of a given reaction with respect to a reactant may be determined in several ways if concentration data are available for isothermal and constant-volume conditions.

For a reaction

$$aA + bB \rightarrow cC \ldots \tag{6.14}$$

n_A is moles of reactant A at any time t, n_{A_0} is moles of reactant A at $t = 0$, n_{B_0} is moles of reactant B at $t = 0$, and x is the amount of conversion, $n_{A_0} - n_A$. Then, for a constant volume system,

$$\frac{dC_A}{dt} = -\frac{dn_A}{dt}\frac{1}{V} = -k_r \frac{(n_{A_0} - x)^a}{V^a} \frac{(n_{B_0} - bx/a)^b}{V^b} \tag{6.15}$$

and

$$\frac{dx}{dt} = k_r\left(\frac{V}{V^n}\right)(n_{A_0} - x)^a\left(n_{B_0} - \frac{b}{a}x\right)^b \tag{6.16}$$

For the special case of a equal to b, and n_{A_0} equal to n_{B_0} (the reaction order n is equal to $a + b$),

$$\frac{dx}{dt} = k_r\left(\frac{V}{V^n}\right)(n_{A_0} - x)^n \tag{6.17}$$

If we take logarithms of both sides of eq 6.17, we get

$$\log \frac{dx}{dt} = n \log(n_{A_0} - x) + \log\left(\frac{k_r V}{V^n}\right) \tag{6.18}$$

A plot of log dx/dt versus $\log(n_{A_0} - x)$ gives the reaction order n as a slope and $(k_r V/V^n)$ as an intercept.

If eq 6.18 is integrated,

$$\left(\frac{1}{n_{A_0} - x}\right)^{n-1} - \left(\frac{1}{n_{A_0}}\right)^{n-1} = (n-1)\left(\frac{k_r V}{V^n}\right)t \qquad (6.19)$$

a plot of $(1/(n_{A_0} - x))^{n-1}$ versus t gives a straight line with a slope $(n-1)(k_r V/V^n)$.

At 50% conversion, the integrated equation takes the general form

$$t_{0.5} = \frac{2^{n-1} - 1}{\left(k_r \dfrac{V}{V^n}\right)(n-1)n_{A_0}^{\,n-1}} \qquad (6.20)$$

for all cases except when reaction order n equals 1. When n equals 1, eq 6.20 becomes

$$t_{0.5} = \frac{\ln\left(\dfrac{2}{V}\right)}{\left(k_r \dfrac{V}{V^n}\right)} \qquad (6.21)$$

Then a plot of $t_{0.5}$ versus log n_{A_0} gives a straight line whose slope is $(1-n)$. The results of eq 6.21 and other techniques are illustrated in Figure 6.4.

These techniques, while simple, give good results as illustrated in Example 6.2, which uses the differential, integrated, and half-time integrated techniques.

EXAMPLE 6.2. CALCULATING THE REACTION ORDER, SPECIFIC REACTION RATE, AND THE HALF-TIME OF THE REACTION

The reaction of a molecule A takes place at constant volume and a temperature of 625 K. Partial pressure data for the reaction as a function of time are as follows:

Time (s)	P_A (10^4 N/m^2)
0	8.45
300	7.89
900	6.89
1500	6.00
2100	5.54
2700	5.15
3300	4.65
3900	4.35
4800	3.93
5400	3.66

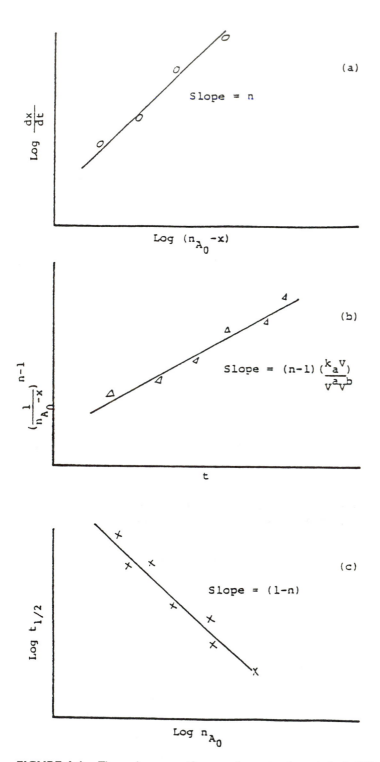

FIGURE 6.4.   The order n, specific rate of reaction k_n, and the half-life of a reaction may be found by plotting time and conversion data as shown in graphs a, b, and c, corresponding to eqs 6.18, 6.19, and 6.20, respectively. Plot a represents the differential approach, plot b, the integrated approach, and plot c, the half-time approach.

The partial pressure, which was initially 1.69×10^5 N/m^2, drops to half its value in 2080 s.

What is the order of the reaction? What is the specific reaction rate? We will use the differential, integrated, and half-time methods.

We first assume that the rate equation is

$$\text{rate} = -\frac{dP_A}{dt} = k_P P_A^n$$

CALCULATING BY DIFFERENTIATION.   We must evaluate $-dP_A/dt$ as a function of P_A to use the differential method. This goal can be accomplished in several ways. We can

1. Plot a curve of P_A versus time and find the slope.
2. Plot a bar-type chart of $\Delta P_A/\Delta t$ versus time and then connect the center points of each bar. The result is a curve that represents dP_A/dt versus t.
3. Differentiate numerically, i.e., using a formula from numerical analysis.

The values for dP_A/dt obtained by any of the three methods are shown in the following table and plotted in Figure 6.5.

$-dP_A/dt$ (N/m^2)	Time (s)
0.0117	0
0.0107	300
0.0084	900
0.0068	1500
0.0052	2100
0.0045	2700
0.0035	3300
0.0029	3900
0.0026	4800

From Figure 6.5,

$$n = \frac{\ln\left(-\dfrac{dP_A}{dt}\right)_1 - \log\left(-\dfrac{dP_A}{dt}\right)_2}{\log P_{A_1} - \log P_{A_2}}$$

$$n = 2$$

$$\frac{-dP_A}{dt} = \frac{2.85 \times 10^{-4}}{\left(\dfrac{N}{m^2}\right)(s)} P_A^2$$

CALCULATING BY INTEGRATION.   If we use the integrated form, eq 6.19, and assume that $n = 2$ the resulting equation is

$$\frac{1}{P_A} - \frac{1}{P_{A_0}} = k_P t$$

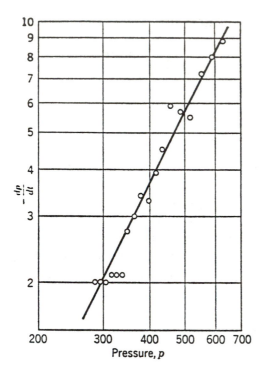

FIGURE 6.5. The reaction rate for the data given in Example 6.2 resulted in this derivative curve for the reaction with pressure in mmHg and time in minutes. (Reproduced with permission from reference 10. Copyright 1989 Stanley Walas.)

A plot of $1/P_A$ versus t gives a straight line (Figure 6.6). The slope is k_P,

$$k_P = \frac{2.85 \times 10^{-4}}{\left(\dfrac{N}{m^2}\right)(s)}$$

CALCULATING BY THE HALF-TIME METHOD.

$$\frac{(t_{0.5})_1}{(t_{0.5})_2} = \left[\frac{(P_{A_0})_2}{(P_{A_0})_1}\right]^{n-1}$$

$$\frac{(4160)}{(2080)} = \left[\frac{(1.69 \times 10^5)}{(8.45 \times 10^4)}\right]^{n-1}$$

$$2 = 2^{n-1}$$

$$n = 2$$

$$k_P = \frac{2^{n-1} - 1}{t_{0.5}(n-1)P_{A_0}} = \frac{2.85 \times 10^{-4}}{\left(\dfrac{N}{m^2}\right)(s)}$$

CALCULATING THE RATE WHEN THE REACTION ORDER IS KNOWN. If the order is known, the rate equation can be directly integrated. For example, for a first-order reaction (constant values)

$$A \rightarrow B + C \tag{6.22}$$

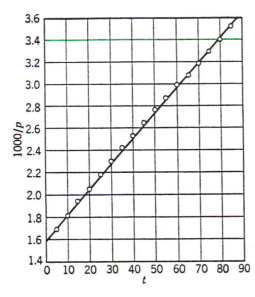

FIGURE 6.6. The integrated rate curve for the data of Example 6.2 has a slope of k_P. (Reproduced with permission from reference 10. Copyright 1989 Stanley Walas.)

the rate is

$$r = \frac{dx}{dt} = k_r(n_{A_0} - x) \tag{6.23}$$

and

$$\ln\left(\frac{n_{A_0} - x_{A_0}}{n_{A_0} - x_A}\right) = k_r(t - t_0) \tag{6.24}$$

For a second-order reaction,

$$2A \xrightarrow{k_r} C + D \tag{6.25}$$

$$-\frac{dC_A}{dt} = k_r C_A^2 \tag{6.26}$$

and

$$\frac{1}{C_A} - \frac{1}{C_{A_0}} = k_r(t - t_0) \tag{6.27}$$

For a reaction of more than one component, where a, b, c, and d indicate stoichiometry,

$$aA + bB \xrightarrow{k_r} cC + dD \tag{6.28}$$

and

$$\frac{dC_A}{dt} = k_r C_A C_B \tag{6.29}$$

then,

$$\frac{1}{C_{B_0} - \frac{b}{a} C_{A_0}} \ln\left[\frac{1}{C_A}\left\{C_{B_0} - \frac{b}{a} C_{A_0} - C_A\right\}\right] - \ln\frac{C_{B_0}}{C_{A_0}} = k_r(t - t_0) \tag{6.30}$$

We shall now consider another example.

EXAMPLE 6.3. CALCULATING THE REACTION RATE AND PERCENT CONVERSION WHEN THE REACTION ORDER IS KNOWN

Five hundred grams of phosphine gas are heated in a tube at 672 °C for 200 s. The dissociation proceeds stoichiometrically as follows:

$$4PH_3 \rightarrow P_4 + 6H_2$$

The reaction is first order. The rate constant, in K, is given by

$$\log_{10} k_r = \frac{-18,963}{T} + 2 \log_{10} T + 12.130$$

We must calculate (a) grams of phosphorus formed in 200 s and (b) time required for 95% decomposition of PH_3. The k_r at 672 °C (945 K) is

$$k_r = 0.010136 \text{ s}^{-1}$$

For the first-order case

$$\frac{n_{A_0} - x_{A_0}}{n_{A_0} - x_A} = e^{k_r(t-t_0)}$$

$$n_A = (n_{A_0} - x_A) = (n_{A_0} - x_{A_0})e^{-k_r(t-t_0)}$$

$$n_A = \left(\frac{500}{34} \text{ g-mol}\right) e^{-0.010136(200)}$$

$$n_a = 1.93 \text{ g-mol}$$

Then,

grams of A = (1.93 g-mol) (34 g-mol) = 65.6 g remaining

For 95% completion,

$$t = \frac{1}{k_r} \ln\left(\frac{n_{A_0} - x_{A_0}}{n_{A_0} - x}\right)$$

then,

$$t = \frac{1}{0.010136 \text{ s}^{-1}} \ln\left(\frac{n_{A_0}}{0.05 n_{A_0}}\right) = 295 \text{ s}$$

In this example, the distinction between reaction stoichiometry and order is shown. A balanced reaction requires a phosphine stoichiometry of 4, while the reaction order is first.

Chemical kinetic calculations can be complicated by reversible, simultaneous, or consecutive reactions. For example, for

$$A + B \underset{k_2}{\overset{k_1}{\rightleftharpoons}} C + D \tag{6.31}$$

the reaction rate expression is

$$r = -\frac{1}{V}\frac{dn_A}{dt} = k_1\left(\frac{n_A}{V}\right)\left(\frac{n_B}{V}\right) - k_2\left(\frac{n_C}{V}\right)\left(\frac{n_D}{V}\right) \tag{6.32}$$

At equilibrium, $r = 0$ and

$$\frac{k_1}{k_2} = K_c = \frac{\left(\dfrac{n_C}{V}\right)\left(\dfrac{n_D}{V}\right)}{\left(\dfrac{n_A}{V}\right)\left(\dfrac{n_B}{V}\right)} \tag{6.33}$$

where K_c is a chemical equilibrium constant based on concentration.

Hence, we can write

$$-\frac{1}{V}\frac{dn_A}{dt} = k_1\left(\frac{n_A n_B}{V^2} - \frac{1}{K_c}\frac{n_C n_D}{V^2}\right) \tag{6.34}$$

or

$$\frac{dx}{dt} = \frac{k_1}{V}\left[(n_{A_0} - x)(n_{B_0} - x) - \frac{(n_{C_0} + x)(n_{D_0} + x)}{K_c}\right] \tag{6.35}$$

Integration of eq 6.35 yields

$$\frac{k_1(t - t_0)}{K_c V} = \frac{1}{q}\ln\left[\frac{(2\alpha x - \beta - q)(2\alpha x_0 - \beta + q)}{(2\alpha x - \beta + q)(2\alpha x_0 - \beta - q)}\right] \tag{6.36}$$

where α is $(K_c - 1)$, β is $(K_c n_{A_0} + K_c n_{B_0} + n_{C_0} + n_{D_0})$, q is $\sqrt{\beta^2 - 4\alpha\gamma}$, and γ is $K_c n_{A_0} n_{B_0} - n_{C_0} n_{D_0}$.

The kinetics calculations are more difficult for consecutive and simultaneous reactions. For the consecutive reactions,

$$2A \overset{k_1}{\longrightarrow} B$$
$$B \overset{k_2}{\longrightarrow} C \tag{6.37}$$

The differential of the rate equation is

$$-\frac{dn_A}{dt} = \frac{k_1 n_A^2}{V} \tag{6.38}$$

Equation 6.38 which can be solved to obtain n_A as a function of time and then substituted in

$$\frac{dn_B}{dt} = \frac{k_1 n_A^2}{V} - k_2 n_B \tag{6.39}$$

to give an equation in terms of only n_B and time.

For simultaneous reactions

$$A \xrightarrow{k_1} C$$

and   (6.40)

$$B \xrightarrow{k_2} C$$

the rate equations are

$$-\frac{dn_A}{dt} = k_1 n_A \tag{6.41a}$$

and

$$-\frac{dn_B}{dt} = k_2 n_B \tag{6.41b}$$

which result in

$$n_A = n_{A_0} e^{-k_1 t} \tag{6.41c}$$

and

$$n_B = n_{B_0} e^{-k_2 t} \tag{6.41d}$$

When the values for n_B and n_A are substituted for C in the kinetic equation,

$$\frac{dn_C}{dt} = k_1 n_A + k_2 n_B \tag{6.42}$$

the resulting equation is

$$\frac{dn_C}{dt} = k_1 n_{A_0} e^{-k_1 t} + k_2 n_{B_0} e^{-k_2 t} \tag{6.43}$$

Tabulations and summaries of solutions for reverse, consecutive, and simultaneous equations are given in references 1–3. Examples 6.4 and 6.5 illustrate how to solve kinetic cases for reversible and simultaneous reactions, respectively. Ingenuity makes difficult situations tractable.

EXAMPLE 6.4. CALCULATING CONVERSION VERSUS TIME AND EQUILIBRIUM CONVERSION FOR A REVERSIBLE REACTION

For the reversible reaction

$$A + B \underset{k_2}{\overset{k_1}{\rightleftarrows}} C + D$$

and with kinetic data

$$k_1 = 7.94 \times 10^{-6} \frac{m^3}{s\text{-kg-mol}}$$

$$k_2 = 2.54 \times 10^{-6} \frac{m^3}{s\text{-kg-mol}}$$

$$C_{A_0} = 4 \frac{kg\text{-mol}}{m^3}$$

$$C_{B_0} = 10.8 \frac{kg\text{-mol}}{m^3}$$

$$C_{C_0} = 18 \frac{kg\text{-mol}}{m^3}$$

$$C_{D_0} = 0$$

we must determine (1) the conversion of A after 2 h and (2) the equilibrium conversion.

If f equals the fraction of A converted, we can rewrite C_A, C_B, C_C, and C_D in terms of the reactants

$$C_A = C_{A_0}(1 - f)$$
$$C_B = C_{B_0} - C_{A_0} f$$
$$C_C = C_{C_0} + C_{A_0} f$$
$$C_D = C_{A_0} f$$

Then,

$$-\frac{dC_A}{dt} = k_1[C_{A_0}(1 - f)][C_{B_0} - C_{A_0} f] - k_2[C_{A_0} f][C_{C_0} + C_{A_0} f] = \frac{df}{dt} C_{A_0}$$

Substituting for known quantities yields

$$\frac{df}{dt} = \frac{k_1}{C_{A_0}} [4 (1 - f)][10.8 - 4f] - \frac{k_2}{C_{A_0}} [4f][18 + 4f] \frac{kg\text{-mol}}{m^3}$$

Then, integrating the above differential from $t = 0$ to any t yields

$$t = 6977 \left[\ln \frac{(f - 7.44)}{(f - 0.552)} \right]_0^{f_t}$$

Then, for 2 h (7200 s), we obtain $f = 0.365$ or 36.5% conversion.
At equilibrium conversion, $dC_A/dt = 0$ and

$$k_1[C_{A_0}(1 - f)][C_{B_0} - C_{A_0}f] = k_2[C_{A_0}f][C_{C_0} + C_{A_0}f]$$

given that k_1, k_2, C_{A_0}, C_{B_0}, and C_{C_0} all are known. Solving, $f_{equilibrium} = 0.55$ or 55% conversion.

EXAMPLE 6.5. WORKING WITH SIMULTANEOUS REACTIONS

We must find the ratios of products produced in the set of simultaneous reactions (constant volume batch reactor),

$$A \xrightarrow{k_1} B$$
$$A \xrightarrow{k_2} C$$
$$A \xrightarrow{k_3} D$$

Then,

$$-\frac{1}{V} \frac{dn_A}{dt} = (k_1 + k_2 + k_3)\frac{n_A}{V}$$

or

$$-\frac{dn_A}{dt} = (k_1 + k_2 + k_3)n_A$$

and

$$\frac{dn_B}{dt} = k_1 n_A$$

$$\frac{dn_C}{dt} = k_2 n_A$$

$$\frac{dn_D}{dt} = k_3 n_A$$

For A,

$$n_A = n_{A_0}e^{-k't}$$

where

$$k' = k_1 + k_2 + k_3$$

and

$$\frac{dn_B}{dt} = k_1(n_{A_0}e^{-k't})$$

$$\frac{dn_C}{dt} = k_2(n_{A_0}e^{-k't})$$

$$\frac{dn_D}{dt} = k_3(n_{A_0}e^{-k't})$$

so that

$$(n_B - n_{B_0}) = \frac{k_1 n_{A_0}}{k'}(1 - e^{-k't})$$

$$(n_C - n_{C_0}) = \frac{k_2 n_{A_0}}{k'}(1 - e^{-k't})$$

and

$$(n_D - n_{D_0}) = \frac{k_3 n_{A_0}}{k'}(1 - e^{-k't})$$

Because the amounts of B, C, and D produced are $(n_B - n_{B_0})$, $(n_C - n_{C_0})$, and $(n_D - n_{D_0})$, then

$$(n_B - n_{B_0})/(n_C - n_{C_0})/(n_D - n_{D_0}) = k_1/k_2/k_3$$

Hence, the ratios of products are in the same ratio as the specific rates of each reaction.

HOMOGENEOUS, ISOTHERMAL, CATALYZED REACTION KINETICS

Homogeneous catalyzed reactions can take place in either the gaseous or liquid phase. The latter is more common.

LIQUID-PHASE REACTIONS—ACID–BASE CATALYSIS. Many liquid-phase homogeneous catalyses involve acid–base catalysis. In such instances, the specific reaction rates can be related to ionization constants by

$$k_r = a(K')^b \tag{6.44}$$

where K' is the ionization constant, and a and b are constants. The value of b varies between 0.3 and 0.9.

GAS-PHASE REACTIONS.   Gas-phase homogeneous catalysis is somewhat rarer than liquid-phase catalysis, but there are a number of examples of such reactions. Table 6.1 shows the catalytic effect of iodine on the gas-phase decomposition of various organic compounds; the activation energy is reduced by the introduction of iodine, which accelerates the reaction.

Generally, the effect of a catalyst is considered to be first order but this is not always true. If, for the conversion of A to R,

$$A \xrightarrow{k_1} R \tag{6.45}$$

and

$$A + \text{catalyst} \xrightarrow{k_2} R \tag{6.46}$$

then,

$$-\frac{dC_A}{dt} = k_1 C_A{}^n + k_2 C_A{}^n(\text{catalyst})^p \tag{6.47}$$

where n and p are reaction orders for A and the catalyst, respectively.

$$-\frac{dC_A}{dt} = [k_1 + k_2(\text{catalyst})^p]C_A{}^n \tag{6.48}$$

and

$$-\frac{dC_A}{dt} = k'C_A{}^n \tag{6.49}$$

Thus, for gas-phase homogeneous catalysis,

$$k' = k_1 + k_2[\text{catalyst}]^n \tag{6.50}$$

which means k' is a function of catalyst concentration.

TABLE 6.1.   Examples of Homogeneous Gas-Phase Reactions Catalyzed by Iodine

Compound Decomposed	Activation Energy, Uncatalyzed Reaction (kcal/g-mol)	Activation Energy, Catalyzed Reaction (kcal/g-mol)
Dipropyl ether	60.5	28.5
Diethyl ether	53.5	34.3
Methylethyl ether	47.0	38.0
Acetaldehyde	45.5	32.5

HOMOGENEOUS, NONISOTHERMAL, UNCATALYZED REACTION KINETICS

In Chapter 1, we saw that temperature change and heat transfer affect the course of a reaction (*see* Figures 1.2 and 1.7). Calculations for reaction systems that involve changing temperatures or heat transfer require knowledge of (1) specific reaction rates as a function of temperature (i.e., Arrhenius or similar relationship), (2) specific heats of reactants and products, (3) heats of reaction, and (4) occurrences of heat transfer.

Solutions for these cases involve coupling chemical kinetics with the first law of thermodynamics. Thus we set the heat transferred equal to the overall change in enthalpy,

$$Q = \Delta H \tag{6.51}$$

Calculations for three basic types of reactors can be handled using this approach: adiabatic, isothermal with heat transfer, and nonisothermal with heat transfer.

ADIABATIC REACTOR.   An adiabatic reactor is one for which heat transferred is assumed to be zero (i.e., the unit is heavily insulated). Hence

$$\Delta H = 0 \tag{6.52}$$

If we add the enthalpies involved in the reaction, we get

$$\Delta H_{\text{reaction}} + \Delta H_{\text{unconverted reactants}} + \Delta H_{\text{products formed}} - \Delta H_{\text{initial reactants}} = 0 \tag{6.53}$$

For reaction

$$A + B \rightarrow C + D \tag{6.54}$$

whose kinetic equation is

$$\text{rate} = \frac{1}{V}\frac{dx}{dt} = \frac{k}{V^2}(n_{A_0} - x)(n_{B_0} - x) \tag{6.55}$$

the enthalpy balance is

$$x(\Delta H_{\text{reaction}})_{T_B} + (n_{A_0} - x)(H_A)_T + (n_{B_0} - x)(H_B)_T + (n_{C_0} + x)(H_C)_T +$$
$$(n_{D_0} + x)(H_D)_T - n_{A_0}(H_A)_{T_0} - n_{B_0}(H_B)_{T_0} - n_{C_0}(H_C)_{T_0} - n_{D_0}(H_D)_{T_0} = 0 \tag{6.56}$$

where T_B is the base temperature for the heat of reaction, T_0 is the initial temperature of the reactants, and T is the final temperature.

For systems with no phase change,

$$x(\Delta H_{\text{reaction}})_{T_B} + (n_{A_0} - x)\int_{T_B}^{T} C_{p_A}dT + (n_{B_0} - x)\int_{T_B}^{T} C_{p_B}dT + (n_{C_0} + x)\int_{T_B}^{T} C_{p_C}dT$$
$$+ (n_{D_0} + x)\int_{T_B}^{T} C_{p_D}dT + \int_{T_0}^{TB}(n_{A_0}C_{p_A} + n_{B_0}C_{p_B} + n_{C_0}C_{p_C} + n_{D_0}C_{p_D})dT = 0 \tag{6.57}$$

If the specific heats (C_p) are relatively constant and if $T_B = T_0$,

$$[(n_{A_0} - x)C_{p_A} + (n_{B_0} - x)C_{p_B} + (n_{C_0} - x)C_{p_C} + (n_{D_0} + x)C_{p_D}](T - T_0)$$
$$+ x(\Delta H_{reaction}) = 0 \quad (6.58)$$

The solution for the adiabatic reactor uses the following procedure:

1. Assume a value of T_1 different from T_0.
2. Calculate x from the appropriate enthalpy balance.
3. Find the value of k_r for a temperature of $(T_1 + T_0)/2$.
4. Calculate $\dfrac{V}{k_r(n_{A_0} - x)(n_{B_0} - x)}$ for the value of T_1.
5. Repeat steps 1 through 4 for T_2 (i.e., get x, get k_r at $(T_2 + T_1)/2$ etc.).
6. When the desired x is finally obtained, find the time t associated with T and x by graphically or numerically integrating the data of step 4 versus x by

$$t = \int_0^x \frac{V}{k_r(n_{A_0} - x)(n_{B_0} - x)} \, dx \quad (6.59)$$

ISOTHERMAL REACTOR WITH HEAT TRANSFER.   The second of the basic cases is the isothermal reactor with heat transfer. Here

$$Q = \Delta H \quad (6.60)$$

However, now $T_0 = T = T_B$ and

$$\left(\frac{\text{heat transferred}}{\text{time}}\right) = (\text{rate of reaction})(\text{heat of reaction})(\text{volume}) \quad (6.61)$$

Again, using the reaction

$$A + B \rightarrow C + D \quad (6.62)$$

the kinetic statement is

$$\text{rate} = \frac{k_r(n_{A_0} - x)(n_{B_0} - x)}{V^2} \quad (6.63)$$

and

$$UA(T_m - T) = (\text{rate})(V)(\Delta H_{reaction})_T \quad (6.64)$$

$$UA(T_m - T) = \frac{k_r(n_{A_0} - x)(n_{B_0} - x)}{V}(\Delta H_{reaction})_T \quad (6.65)$$

In eqs 6.64 and 6.65, U is the overall heat transfer coefficient, A is the heat transfer surface area, and T_m is the temperature of the heat transfer fluid.

NONADIABATIC, NONISOTHERMAL REACTOR. The third of the cases involves a reactor in which temperature changes and there is heat transfer. This reactor is called a programmed reactor. For the system,

$$\text{heat transferred} = \Delta H_{\text{reaction}} + \Delta H_{\text{unconverted reactants}} + \Delta H_{\text{products formed}}$$
$$- \Delta H_{\text{initial reactants}} \qquad (6.66)$$

This is a more complex problem than the other two cases and will be illustrated with an example.

EXAMPLE 6.6. HOMOGENEOUS, NONISOTHERMAL, NONADIABATIC REACTION CALCULATIONS—A TRIAL-AND-ERROR TECHNIQUE. We will consider an ideal gas decomposition

$$D \rightarrow E + F$$

that takes place in a nonisothermal reactor within the range 327–338 K with up to 90% conversion. The water used to cool the reactor remains at 300 K. The overall heat transfer coefficient U is 5.26 J/s-m^2-K. How large is the amount of heat transfer surface needed?

The heat of reaction of 333 K is -5.8×10^6 J/kg-mol. Initial conditions are 330 K, 5.07×10^5 N/m^2 pressure, and a volume of 0.2475 m^3. Specific heats for D, E, and F are 1.252×10^5, 1.043×10^5, and 1.043×10^5 J/kg-mol-K, respectively. The Arrhenius relation for the specific reaction rate is

$$k = 2.2 \times 10^6 \, e^{-E/RT} \text{s}^{-1}$$

where E is 6.26×10^7 J/kg-mol.

To solve this problem, we must set up an overall enthalpy balance for the reactor. This balance will include the heat of reaction, the enthalpy changes of reactant and products, and the heat transferred to the cooling water.

$$\text{heat transferred} = \Delta H_{\text{unconverted reactant}} + \Delta H_{\text{products formed}}$$
$$- \Delta H_{\text{initial reactants}} + \Delta H_{\text{reaction}}$$

In symbolic form, for any time interval during which an amount x of reactant D is converted

$$UA(T_m - T)dt = (n_{D_0}C_{p_D} + x\Delta C_p)dT + (\Delta H_r)dx$$

ΔH_r is the heat of reaction in joules per kilogram-mole, U is the overall heat transfer coefficient in joules per second-square meter-kelvin, A is the heat transfer surface in square meters, T_m is the temperature of the cooling water in kelvins, n_{D_0} is the original number of moles of D, and ΔC_p is $(C_{p_F} + C_{p_E} - C_{p_D})$ in joules per kilogram-mole-kelvin.

Rearranging the above equation and integrating from T_0, the original temperature at $t = 0$, results in

$$(\Delta H_r)x + (n_{D_0}C_{p_D} + x\Delta C_p)(T - T_0) = \int_0^x UA(T_m - T)dt$$

Because $rV dt = dx$ (eq 6.1)

$$(\Delta H_r)x + (n_{D_0}C_{p_D} + x\Delta C_p)(T - T_0) = \int_0^x \frac{UA(T_m - T)}{rV}\, dx$$

In this equation, A, ΔC_p, T, x, t, and n_{D_0} are all unknowns which, except for n_{D_0}, must be calculated by trial and error.

We can calculate n_{D_0} easily because we are working with an ideal gas.

$$n_{D_0} = \frac{PV}{RT} = \frac{\left(5.06 \times 10^5\, \frac{\text{N}}{\text{m}^3}\right)(0.2475\text{ m}^3)}{\left(\dfrac{8.314 \times 10^3\, \frac{\text{N}}{\text{m}^2}\, \text{m}^3}{\text{kg-mol-K}}\right)(333\text{ K})}$$

$$n_{D_0} = 0.0455\text{ kg-mol}$$

To start the trial-and-error calculation, we assume isothermal operation to determine a value for A. For an isothermal reaction

$$t = \frac{1}{3.33 \times 10^{-4}}\ln\frac{0.1}{(0.1 - 0.09)} = 6920\text{ s}$$

Then, for $t = 6920$ s and isothermal conditions,

$$UA(T_m - T) = (\text{rate})(V)(\Delta H_{\text{reaction}})_T$$

$$A = \frac{Q}{tU\Delta T} = \frac{x\Delta H_{\text{reaction}}}{tU\Delta T}$$

$$A = \frac{(0.041\text{ kg-mol})(5.8 \times 10^6\text{ J/kg-mol-K})}{(6920\text{ s})(5.26\text{ J/s-m}^2\text{-K})(333 - 300\text{ K})}$$

$$A = 0.061\text{ m}^2$$

For the nonisothermal case, we will assume that a somewhat larger surface area (0.0929 m^2) is available for heat transfer because more surface will be needed. Then, returning to our equation, seeing that $rV = k(n_{D_0} - x)$, and multiplying both sides by minus 1,

$$-(\Delta H_{\text{reaction}})x + (n_{D_0}C_{p_0} + x\Delta C_p)(T - T_0) = \int_0^x \frac{UA(T_m - T)}{k(n_{D_0} - x)}\, dx$$

Substituting known values in the equation,

$$(5.8 \times 10^6)x + (5.7 \times 10^3 + 8.34 \times 10^4\, x)(T - 333)$$

$$= \int_0^x \frac{5.26(0.0929)(T - 300)}{k(0.0455 - x)}\, dx$$

This is an equation with two unknowns (k is a dependent variable of T), which can be solved by trial and error.

If we start with x equal to 0.00455 ($x_1 = 0.00455$), we can let T_1 be an assumed temperature (335 K), and check the equation (integrating graphically and taking the T for k to be $(T_0 + T_1)/2$). When checked, T_1 should be 334.9 K. Now we can repeat the calculation with a change of x until the conversion converges to $x = 0.04095$ as suggested by the table below.

x (kg-mol)	T (K)
0	333
0.00455	334.9
0.01365	338
0.02220	338.5
0.03185	334.9
0.04095	318

Because the temperature (318 K) at $x = 0.04095$ (90% conversion) is less than the specified lower temperature limit, we would repeat the calculations assuming a smaller surface area until we can achieve 90% conversion at a temperature above 327 K. The actual area will be 0.088 m^2 and the time, 4690 s.

EXAMPLE 6.7. ADIABATIC, ISOTHERMAL REACTION CALCULATIONS. For the system of the preceding example, we must compare the conversion-time behavior for the adiabatic and isothermal cases if the initial temperature is 333 K.

In the adiabatic case,

$$\Delta H = 0$$

and

$$[(n_{D_0} - x)C_{p_D} + xC_{p_E} + xC_{p_F}](T - T_0) - x(\Delta H_{\text{reaction}}) = 0$$

Substituting values where possible,

$$\left[(0.0455 - x \text{ kg-mol})\left(1.25 \times 10^5 \frac{J}{\text{kg-mol-K}}\right) + (2x \text{ kg-mol})\right.$$

$$\left.\left(1.043 \times 10^5 \frac{J}{\text{kg-mol-K}}\right)\right](T - T_0) - \left(5.8 \times 10^6 \frac{J}{\text{kg-mol}}\right)x \text{ kg-mol} = 0$$

The kinetic statement gives

$$t = \int_0^x \frac{dx}{k(0.0455 - x)}$$

If we carry out the method described earlier for the adiabatic case, we obtain

x (kg-mol)	T (K)
0	333
0.009	341
0.018	348
0.027	352
0.036	358
0.041	359

Time t versus the amount of conversion x can be calculated from the above kinetic statement by using a k_r value based on the average temperature over the interval.

The t vs. x data for the isothermal case is obtained directly from the integrated rate expression using the k_r at 333 K:

$$x = n_{D_0}(1 - e^{-k_r t})$$

Values of t vs. x are in the following table:

x (kg-mol)	T (K)	t, Adiabatic (s)	t, Isothermal (s)
0	333	0	0
0.009	341	511	670
0.018	348	945	1540
0.027	352	1360	2750
0.036	358	1900	4825
0.041	359	2335	6900

From this example, it appears that an adiabatic reactor would be the reactor of choice. This, however, is possible only if the following conditions are met:

1. Heat of reaction is small.
2. Initial temperature can be adjusted so that the changes will not take the system out of the reaction range.
3. Heat transfer capacity of the equipment, a solvent, or inert materials can moderate the temperature.
4. An inert material can be used to moderate the temperature by its vaporization or condensation.

Caution must be exercised because an adiabatic reactor's temperature can drop so low that it inhibits or stops the reaction or rise so sharply and rapidly that the reaction rate becomes uncontrolled. Heat transfer or heat flux must be regulated by computer programming.

Solutions for isothermal, adiabatic, and programmed reactors require combining the first law of thermodynamics with the chemical kinetics of the system. The enthalpy portion of the energy balance is familiar to chemists; only the heat flux and heat transfer are new.

HETEROGENEOUS, ISOTHERMAL, UNCATALYZED CHEMICAL KINETICS AND THE INFLUENCE OF MASS TRANSFER

Mass transfer is important in both uncatalyzed and catalyzed heterogeneous reaction systems because the reactants must migrate to a phase boundary before they can react. Physical factors are also important and can be rate determining. The specific factors that are important include (1) the size of the boundary between phases, (2) the rate of diffusion of reactants to and past the phase boundary, and (3) the rate of diffusion of products away from the reaction zone.

Examples of uncatalyzed heterogeneous reactions are listed in Table 6.2.

Experimental studies can show whether chemical reaction or mass transfer is rate determining. A large temperature effect, per the Arrhenius equation, indicates the predominance of chemical reaction. However, significant shifts in the products formed, when there are changes in flow or interfacial areas, indicate that mass transfer predominates. Some examples of reactions in which mass transfer is rate determining include the combustion of carbon in air, ion-exchange systems, and the production of polymer by interfacial polycondensation. In most ion-exchange systems, the chemical reaction itself is relatively fast, which means that diffusion controls the process.

Gas absorption in a liquid, accompanied by a chemical reaction, has been one of the more widely studied heterogeneous, uncatalyzed reaction systems. Several variations of the gas–liquid system have been considered, including absorption with slow reaction, absorption with fast reaction, different orders of reaction, and reversible cases.

If a slow reaction takes place between an absorbed gas molecule D with a mol-

TABLE 6.2. Examples of Heterogeneous Uncatalyzed Reactions

Phases Involved	Reaction
Gas–solid	Combustion of coal
	Production of hydrogen by steam contact with iron
	Nitriding of steel
Gas–liquid	Production of nitric acid by absorption of nitric oxide in water
	Hydrogenation of vegetables oils
	Removal of hydrogen sulfide gas by aqueous ethanolamines
Liquid–solid	Cyaniding of steel
	Hydration of lime
	Ion exchange
Liquid–liquid	Reaction of petroleum fractions with sulfuric acid
	Nitration of organics
Solid–solid	Cement production
	Production of boron carbide from carbon and boron oxide

ecule E (contained in a liquid) in a liquid film, then according to the equation of continuity of species (*see* Chapter 5),

$$D_{DE} \frac{d^2 C_D}{dx^2} = k_r C_D \qquad (6.67)$$

where D_{DE} is the diffusivity of D in liquid E, and x is the direction in which D diffuses (Figure 6.7). The reaction is written as first order because there is a large excess of E.

Solution of eq 6.67 by applying appropriate boundary conditions leads to a dimensionless group, the Hatta number,

$$Ha = \frac{\text{mass transfer with chemical reaction}}{\text{mass transfer without chemical reaction}} \qquad (6.68a)$$

The reaction rate with diffusion r_d is then

$$r_d = k_L^0 (Ha)(\Delta C_D) = k_L \Delta C_D \qquad (6.68b)$$

where k_L^0 is the mass transfer coefficient without chemical reaction and k_L is the mass transfer coefficient with reaction.

$$Ha = \frac{b_1 \delta}{\tanh b_1 \delta} \qquad (6.69)$$

where δ is the thickness of the liquid film and b_1 is $(k_r/D')^{0.5}$.

A correlation between the mass transfer coefficient and chemical reaction is

$$\frac{k_L \delta}{D'} = 0.015 \left(\frac{L}{a_v \mu} \right)^{0.66} \left(\frac{\mu}{\rho D'} \right)^{0.33} (Ha) \qquad (6.70)$$

where a_v is the interfacial area-to-volume ratio, L is the mass velocity [mass/(time)(area)], D' is the diffusivity in liquid, and k_L is the mass transfer coefficient.

$$\delta = \left(\frac{\mu^2}{g\rho^2} \right)^{0.33} \qquad (6.71)$$

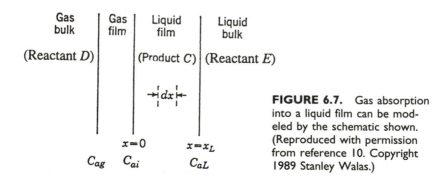

| Gas bulk | Gas film | Liquid film | Liquid bulk |

(Reactant D)   (Product C) (Reactant E)

$\rightarrow |dx| \leftarrow$

$x = 0$ $x = x_L$

C_{ag} C_{ai} C_{aL}

FIGURE 6.7. Gas absorption into a liquid film can be modeled by the schematic shown. (Reproduced with permission from reference 10. Copyright 1989 Stanley Walas.)

The b value for a first-, second-, or third-order reaction is given, respectively, by

$$b_1 = \sqrt{\frac{k_{r_1}}{D_{DE}}} \tag{6.72}$$

$$b_2 = \sqrt{\frac{k_{r_2}}{D_{DE}}} \tag{6.73}$$

$$b_3 = \sqrt{\frac{k_{r_3}}{D_{DE}}} \tag{6.74}$$

for which the k_{r_1}, k_{r_2}, and k_{r_3} values are specific reaction rates for first-, second-, and third-order cases.

A number of other solutions have been prepared for other cases. An example is given (Figure 6.8) for a reversible first-order reaction.

HETEROGENEOUS, ISOTHERMAL, CATALYZED REACTIONS—SURFACE REACTION DEPENDENCY

Catalyzed, heterogeneous reactions are of great industrial importance. They are the principal methods of synthesis in the petroleum and the petrochemical industries. Most heterogeneous, catalyzed systems use solid catalysts because these catalysts are thermally stable and easily separated from the reaction mix.

The mechanism of solid catalysis can be described as a series of steps: (1) diffusion of reactants to the catalyst surface, (2) adsorption of reactants on the surface, (3) reaction on the surface, (4) desorption of productions from the surface, and (5) diffusion of products into the fluid.

If we attempted to link all of these steps together, we would have a difficult problem to analyze. Fortunately, one or more of the steps predominate, which simplifies determining the rate of reaction.

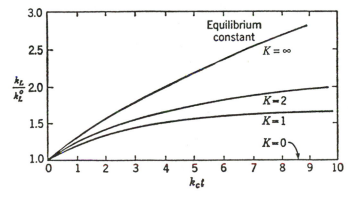

FIGURE 6.8. Mass transfer coefficient ratio for first-order reversible reactions, where K is the equilibrium constant, k_c is the reaction rate constant, and k_L/k_L^0 is the mass transfer coefficient correction. (Reproduced from reference 11. Copyright 1953 American Chemical Society.)

Adsorption, step 2, figures prominently in many solid-catalysis reactions so that many of the rate expressions are related to the equilibrium constant for adsorption, derived as follows:

$$r_1 = \text{rate of adsorption} = k_1 P_A (1 - \theta) \tag{6.75}$$

where P_A is the partial pressure of material A adsorbed, k_1 is the specific rate of adsorption, and θ is the fraction of surface covered.

For desorption, the derivation results in

$$r' = \text{rate of desorption} = k'_1 \theta \tag{6.76}$$

where k'_1 is the specific desorption rate. Then,

$$K_z = \frac{k_1}{k'_1} \tag{6.77}$$

where K_z is the adsorption equilibrium constant for compound z.

The most important steps in the solid catalyst mechanism are adsorption, reaction, and desorption; diffusion is not usually rate controlling. There are two basic cases considered when the adsorption–reaction–desorption scheme holds: (1) equilibrium adsorption with the surface reaction controlling the rate and (2) rapid surface reaction with adsorption controlling the rate. The analysis of each case leads to equations that can be written as

$$\text{rate} = \frac{k' \ (\text{driving force term})}{(\text{adsorption term})} \tag{6.78}$$

Table 6.3 lists the driving force and adsorption terms for a number of cases. In the table, K^* is defined as

$$K^* = \frac{\left(\begin{array}{c}\text{products of}\\ \text{adsorption equilibrium}\\ \text{constants for reactants}\end{array}\right)\left(\begin{array}{c}\text{chemical}\\ \text{equilibrium}\\ \text{constant}\end{array}\right)}{\left(\begin{array}{c}\text{products of adsorption}\\ \text{equilibrium constants for}\\ \text{products}\end{array}\right)} \tag{6.79}$$

An example of K^* for case 2 of Table 6.3 is

$$K^* = \frac{K_A K_{eq}}{K_M} \tag{6.80}$$

similarly, for case 4,

$$K^* = \frac{K_A K_B K_{eq}}{K_M K_N} \tag{6.81}$$

TABLE 6.3. Rate Expressions for Reactions Catalyzed with Solid Catalysts

Reaction	Surface Reaction Controlling		Adsorption Rate Controlling	
	Driving Force	Adsorption Term	Driving Force	Adsorption Term
1. $A \rightarrow M + N$	P_A	$(1 + K_A P_A + K_M P_M + K_N P_N)$	P_A	$\left(1 + \dfrac{K_A P_M P_N}{K^{**}} + K_M P_M + K_N P_N\right)$
2. $A \rightleftharpoons M$	$P_A - \dfrac{P_M}{K^*}$	$(1 + K_A P_A + K_M P_M)$	$P_A - \dfrac{P_M}{K^{**}}$	$\left(1 + \dfrac{K_A P_M}{K^{**}} + K_M P_M\right)$
[a]3. $A + B \longrightarrow M + N$	$P_A P_B$	$(1 + K_A P_A + K_B P_B + K_M P_M + K_N P_N)^2$	P_A	$\left(1 + \dfrac{K_A P_M P_N}{K^{**} P_B} + K_M P_M + K_N P_N\right)$
4. $A + B \rightleftharpoons M + N$	$P_A P_B - \dfrac{P_M P_N}{K^*}$	$(1 + K_A P_A + K_B P_B + K_M P_M + K_N P_N)^2$	$P_A - \dfrac{P_M P_N}{K^{**} P_B}$	$\left(1 + \dfrac{K_A P_M P_N}{K^{**} P_B} + K_B P_B + K_M P_M + K_N P_N\right)$
[b]5. $A_2 + B \rightleftharpoons M + N$	$P_A P_B - \dfrac{P_M P_N}{K^*}$	$(1 + \sqrt{K_A P_A} + K_B P_B + K_M P_M + K_N P_N)^3$	$P_A - \dfrac{P_M P_N}{K^{**} P_B}$	$\left(1 + \sqrt{\dfrac{K_A P_M P_N}{K^{**} P_B}} + K_B P_B + K_M P_M + K_N P_N\right)^2$

[a]In Case 3 (adsorption controlling), unadsorbed A reacts.
[b]In Case 5, adsorbed A_2 dissociates.

K^{**} is defined as

$$K^{**} = \frac{\left(\begin{array}{c}\text{products of}\\\text{adsorption equilibrium}\\\text{constants for reactants}\end{array}\right)\left(\begin{array}{c}\text{overall equilibrium}\\\text{constant for surface}\\\text{reaction}\end{array}\right)}{\left(\begin{array}{c}\text{products of adsorption}\\\text{equilibrium constants for}\\\text{products}\end{array}\right)} \tag{6.82}$$

For case 4 an example of a K^{**} is

$$K^{**} = \frac{K_A K_B K'}{K_M K_N} \tag{6.83}$$

In eq 6.83, K' is the overall equilibrium constant for the surface reaction. The rate expressions for gas–solid catalyst systems become complex.

In many instances, the rate can be expressed simply in standard kinetic forms, as for example, carbon monoxide (CO) reacting with chlorine (Cl_2) over charcoal to yield phosgene ($COCl_2$). The rate expression derived from a step-by-step treatment of the adsorption–reaction–desorption process is

$$r = \frac{k K_{CO} K_{Cl_2} P_{CO} P_{Cl_2}}{(1 + K_{Cl_2} P_{Cl_2} + K_{COCl_2} P_{COCl_2})^2} \tag{6.84}$$

The empirically derived expression for the same reaction is

$$r = k P_{CO}(P_{Cl_2})^{0.5} \tag{6.85}$$

Although eqs 6.84 and 6.85 appear to be different, they are not as dissimilar as they seem. The numerator of eq 6.84, $k K_{CO} K_{Cl_2} P_{CO} P_{Cl_2}$, can be combined in a single k term. Also, the partial pressure term for carbon monoxide is the same for both equations. Finally, the partial pressure term for chloride in both equations is raised to a fractional power.

Another reaction for which the rate expression can be simplified is the reaction between methane and sulfur in the presence of a silica gel catalyst. The simplified expression is

$$r = k P_{CH_4} P_{S_2} \tag{6.86}$$

Heterogeneous reactions have complicated rate equations because processes other than the reactions (diffusion or adsorption) are involved. However, it is possible to obtain simpler expressions.

CHEMICAL REACTOR DESIGN

Designing chemical reactors requires knowledge about reaction kinetics, fluid flow, heat transfer, and mass transfer.

The simplest type of reactor is the batch or nonflow system. In this system, reactants are charged into the reactor and held for a specified period of time to yield the desired products. Nonflow reactors are particularly appropriate for laboratory studies because, on this scale, they have a large amount of surface area per unit volume, which allows for excellent temperature control. They are also simple to operate. Batch reactors, however, are inappropriate for processes including those with short reaction times, gaseous reactants, high production rates, and process control problems.

For these processes, a flow reactor is used. Flow reactors can take the form of tubular flow, steady-state tank flow, and semibatch (these reactors were briefly discussed in Chapter 1; *see* Figures 1.10–1.16). The residence time governs the degree of conversion.

Among the types of flow reactors, the fixed-bed (Figure 6.9), moving-bed (Figure 6.10), and fluidized-bed (Figure 6.11) are widely used for heterogeneous catalytic reactions. Their design is complex, requiring simultaneous solution of the differential equations for energy and continuity of species.

TUBULAR FLOW REACTORS

The tubular flow reactor is a widely used unit with several advantages: it is easy to control, saves labor, is mechanically simple, is adaptable to heat transfer, produces product of consistent quality, and has high throughput capacity.

VOLUME DETERMINATION—RESIDENCE TIME AND SPACE VELOCITY

Time and temperature are the principal parameters for chemical reaction in general. Time is a function of distance traveled for a tubular flow reactor. Hence,

$$\theta = \text{residence time} = \frac{V_r}{w/\rho} \tag{6.87}$$

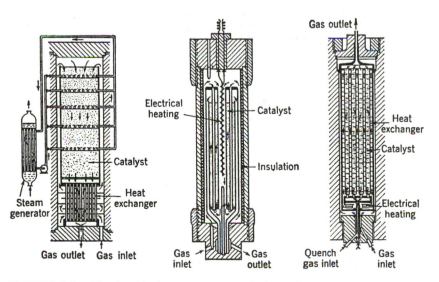

FIGURE 6.9. The fixed-bed reactor, a type of tubular flow reactor, consists of an unmoving bed of solid catalyst with gas flowing over it. (Reproduced with permission from reference 12. Copyright 1950 Carl Hansen Verlag.)

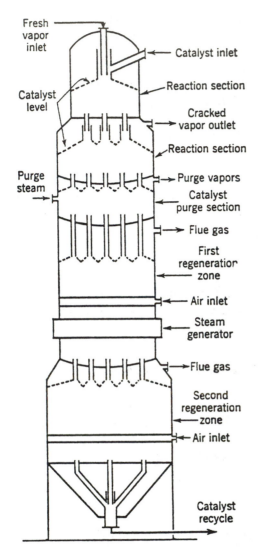

FIGURE 6.10. In a moving-bed reactor, both the catalyst and the gas flow through the reactor. (Reproduced with permission from reference 10. Copyright 1989 Stanley Walas.)

where w is mass flow rate, ρ is the density, and V_r is the reactor volume. The inverse of θ is called the space velocity, SV,

$$\text{SV} = 1/\theta \tag{6.88}$$

For a reactor with plug flow (ideal flow, i.e., the fluid's velocity profile is flat across the tube cross section), by mass balance

$$r_A dV_r = W df_A \tag{6.89}$$

where r_A is the reaction rate for A, f_A is the fraction reacted $(n_{A0} - n_A)/n_{A0}$, and W is the molal feed rate. Then,

$$\frac{V_r}{W} = \int_0^{V_r} \frac{dV_r}{W} = \int_{f_0}^{f} \frac{df_A}{r_A} \tag{6.90}$$

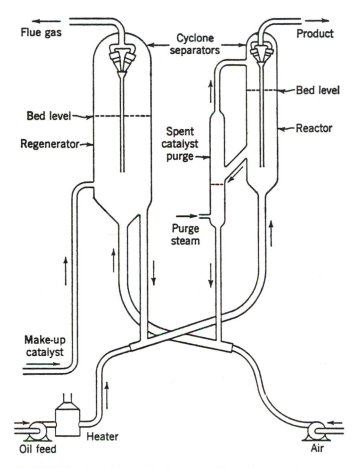

FIGURE 6.11. Fluidized-bed reactor. (Reproduced with permission from reference 10. Copyright 1989 Stanley Walas.)

and

$$dt = \frac{n_{A_0} df_A}{r_A V_r} \tag{6.91}$$

For a gaseous system, the volume can vary because of a change in the number of moles due to reaction. Thus, the number of moles present at any time is $(n_{t_0} + \gamma n_{A_0} f_A)$ where γ is the increase in number of moles per mole of A converted.

If it is assumed that the gas is ideal, then

$$dt = \frac{n_{A_0} df_A}{r_A V_r} = \frac{n_{A_0} df_A}{\left(\dfrac{(n_{t_0} + \gamma n_{A_0} f_A) RT}{P}\right) r_A} \tag{6.92}$$

Applying the foregoing to the case of the irreversible first-order reaction,

$$r_A = \frac{k_1 n_{A_0}(1 - f_A)}{V_r} = \frac{k_1 n_{A_0}(1 - f_A)}{\left(\dfrac{(n_{t_0} + \gamma n_{A_0} f_A) RT}{P}\right)} \tag{6.93}$$

Then,

$$\frac{V_r}{W} = \int_0^{f_A} \frac{df_A}{r_A} = \int_0^{f_A} \left(\frac{RT}{Pk_1}\right) \frac{(n_{t_0} + \gamma n_{A_0} f_A) df_A}{n_{A_0}(1 - f_A)} \tag{6.94}$$

and, if pressure and temperature are constant, then

$$\frac{V_r}{W} = \frac{RT}{Pk_1} \int_0^{f_A} \frac{(n_{t_0} + \gamma n_{A_0} f_A) df_A}{n_{A_0}(1 - f_A)} \tag{6.95}$$

so that

$$V_r = \frac{WRT}{Pk_1}\left[\left(\frac{n_{t_0} + \gamma n_{A_0}}{n_{A_0}}\right) \ln \frac{1}{(1 - f_A)} - \gamma f_A\right] \tag{6.96}$$

Solutions for other kinetic cases can be set up and solved in a similar manner. We now consider some examples, the first of which is a reworking of the phosphine decomposition problem of Example 6.3 for a flow system.

EXAMPLE 6.8. CALCULATING TUBULAR REACTOR VOLUME FOR A FIRST-ORDER, IRREVERSIBLE REACTION

The phosphine reaction is

$$4PH_3 \xrightarrow{k_1} P_4 + 6H_2$$

The k_1 at 920 K is 2.78×10^{-3} s^{-1}, a first-order reaction constant. If the pressure is 4.66×10^5 N/m^2, what volume of tubular reactor is needed for 80% conversion of a feed of 1.82 kg-mol of phosphine per hour?

For the first-order case, we can use the solved eq 6.96,

$$V_r = \frac{WRT}{Pk_1}\left[\left(\frac{n_{t_0} + \gamma n_{A_0}}{n_{A_0}}\right) \ln \frac{1}{(1 - f_A)} - \gamma f_A\right]$$

Because $\gamma = (7 - 4)/4 = 3/4$,

$$V_r = \left[\frac{\left(\dfrac{(1.82 \text{ kg-mol})}{3600 \text{ s}}\right)\left(8314.4 \dfrac{m^3\text{-}N/m^2}{\text{kg-mol-K}}\right)(920 \text{ K})}{(4.66 \times 10^5 \text{ N/m}^2)(2.68 \times 10^{-3} \text{ s}^{-1})}\right] \times$$

$$\left[\left(\frac{4 + 4(0.75)}{4}\right) \ln \frac{1}{(1 - 0.8)} - (0.75)0.8\right]$$

$$V_r = 10.57 \text{ m}^3$$

In more complex kinetic analyses, it is frequently necessary to use numerical or graphical solutions.

EXAMPLE 6.9. CALCULATING THE REACTOR VOLUME NEEDED FOR A REVERSIBLE REACTION WITH INTEGRATION OR GRAPHING

For the reversible reaction

$$A \underset{k_2}{\overset{k_1}{\rightleftharpoons}} 2B$$

at 833 K and 5.05×10^5 N/m^2, we must calculate the reactor volume and space velocity if the feed contains 30 mol% A (70% inert materials) and the feed rate is 34.1 kg-mol/h. Also, $k_1 = 1.6$ s^{-1} and $K_{eq} = 0.0055$. The desired conversion is 75% of equilibrium.

$$n_t = n_{t_0} + n_{A_0} f_A \gamma$$

and

$$\gamma = \frac{2 - 1}{1} = 1$$

so that $n_t = n_{t_0} + n_{A_0} f_A$. Taking a basis of 1 mol of feed

$$V = \frac{n_t R T}{P} = \frac{(1 + 0.3\, f_A)\left(8314.4\ \dfrac{\text{m}^3\text{-N/m}^2}{\text{kg-mol-K}}\right)(833\ \text{K})}{(5.05 \times 10^5\ \text{N/m}^2)}$$

so that

$$V = (1 + 0.3 f_A)(13.72)\ \text{m}^3$$

At equilibrium,

$$K_c = 0.0055 = \frac{C_B^2}{C_A} = \frac{n_B^2}{V n_A}$$

and

$$0.0055 = \frac{(2 n_{A_0} f_A)^2}{(n_{A_0} - n_{A_0} f_A)(1 + 0.3 f_A)(13.72)}$$

Then, f_A at equilibrium is 2/3.

The kinetic statement for the reaction is

$$r_A = k_1 \left[\frac{(n_{A_0} - n_{A_0} f_A)}{V} - \frac{1}{K_{eq}} \frac{(2 n_{A_0} f_A)^2}{V^2} \right]$$

Combining this with the definition for V and substituting in the equation

$$\frac{V_r}{W} = \int_0^{f_A} \frac{df_A}{r_A}$$

yields

$$\frac{V_r}{W} = \int_0^{f_A} \frac{df_A}{k_1 \left[\frac{(0.3 - 0.3 \, f_A)}{(1 + 0.3 \, f_A)(13.72)} - \frac{1}{0.0055} \frac{(0.6 f_A)^2}{(13.72)^2 (1 + 0.3 \, f_A)^2} \right]}$$

When the above equation is integrated numerically or graphically, the result is

$$\frac{V_r}{W} = 8.42 \; \frac{m^3\text{-s}}{\text{kg-mol}}$$

$$V_r = \left(\frac{34.1 \text{ kg-mol}}{3600 \text{ s}} \right) \left(8.42 \; \frac{m^3\text{-s}}{\text{kg-mol}} \right) = 0.0797 \text{ m}^3$$

Then the space velocity (SV) is

$$SV = \frac{\text{feed (standard volume)}}{\text{reactor volume}}$$

$$SV = \frac{\left(\frac{34.1 \text{ kg-mol}}{3600 \text{ s}} \right) \left(22.4 \; \frac{m^3}{\text{kg-mol}} \right)}{0.0797 \text{ m}^3} = 2.66 \text{ s}^{-1}$$

FRICTION CONSIDERATIONS

In a flow reactor, fluid friction losses can affect the total pressure and must be taken into account during the design. We need to develop a relation between the change in pressure due to friction and the reaction rate. To begin, we can write the relation between pressure drop and friction factor f as

$$\frac{dP}{dL} = \frac{2 \, f \rho V^2}{g_c D} \tag{6.97}$$

For turbulent flow, there are empirical relations for f in terms of the Reynolds number. A typical relation is

$$f = \frac{0.079}{\text{Re}^{0.25}} \tag{6.98}$$

Because

$$w = \frac{\pi}{4} D^2 \rho \overline{V} \tag{6.99}$$

and

$$\frac{dP}{dL} = \frac{2 \, f \rho}{g_c D} \left(\frac{4w}{\pi D^2 \rho} \right)^2 \tag{6.100}$$

then,

$$\frac{dP}{dL} = \frac{2\rho}{g_c D} \left(\frac{4w}{\pi D^2 \rho}\right)^2 \frac{0.079}{Re^{0.25}} \tag{6.101}$$

But the Reynolds number can also be expressed as

$$Re = \frac{4w}{\pi D \mu} \tag{6.102}$$

so that

$$dP = 0.241 \left(\frac{w^{7/4} \mu^{1/4}}{g_c D^{19/4} \rho}\right) dL \tag{6.103}$$

The density can be expressed as

$$\rho = \frac{PM}{RT} = \frac{PM_0 n_{t_0}}{RT n_t} \tag{6.104}$$

$$\rho = \frac{PM_0 n_{t_0}}{RT(n_{t_0} + n_{A_0} f_A \gamma)} \tag{6.105}$$

$$r_A dV_r = r_A(\frac{\pi}{4} D^2 \, dL) = \frac{w}{M_0} n_{A_0} df_A \tag{6.106}$$

and

$$dL = \frac{4}{\pi} \frac{w n_{A_0} df_A}{M_0 D^2 r_A} \tag{6.107}$$

Substituting for both dL and ρ in eq 6.103 yields

$$P dP = \frac{0.307 w^{11/4} \mu^{1/4} RT n_{A_0}(n_{t_0} + n_{A_0} f_A \gamma) df_A}{g_c M_0^2 D^{27/4} n_{t_0} r_A} \tag{6.108}$$

Thus we have derived an equation showing that the rate of reaction will be some function of f_A and P,

$$r = \phi(f_A, P) \tag{6.109}$$

If we substitute an appropriate rate statement in eq 6.108, we obtain a relationship between P and an integral involving a number of terms and df_A.

As an illustration, we shall consider the first-order case

$$r_A = \frac{k_1 n_{A_0}(1 - f_A)P}{(n_{t_0} + n_{A_0} f_A \gamma)RT} \tag{6.110}$$

Then,

$$\int_{P_0}^{P} P^2 \, dP = \frac{0.307 \; w^{11/4} \mu^{1/4} (RT)^2 n_{A_0}}{g_c M_0^2 k_1 n_{t_0} n_{A_0} D^{27/4}} \int_0^{f_A} \frac{(n_{t_0} + n_{A_0} f_A \gamma)^2 \, df_A}{(1 - f_A)} \quad (6.111)$$

This means that P can be found as a function of f_A. After this is done, the reactor volume can be calculated from eq 6.106.

A technique that simplifies the process of finding the reactor volume consists of first calculating the reactor length (i.e., get V_r and find L) as if there were no change of pressure due to friction. Next, we compute the pressure drop for the estimated reactor length using the friction factor relationship; this relationship gives an indication of the exit pressure. If the change in pressure is small, an average value for pressure can be used as in the following equation for a first-order reaction,

$$\frac{V_r}{W} = \frac{P_{\text{average}}}{RTk_1} \int_0^{f_A} \frac{(n_{t_0} + n_{A_0} f_A \gamma)^2 \, df_A}{(1 - f_A)} \quad (6.112)$$

EFFECT OF LAMINAR FLOW ON REACTOR DESIGN

If flow is highly turbulent (has a blunt velocity profile), then ideal flow can be assumed with good results. However, if flow is laminar and nonideal, then

$$v = 2\overline{V} \left[1 - \left(\frac{r}{R} \right)^2 \right] \quad (6.113)$$

where v is velocity at radius r, $\overline{V}$ is the mean velocity, and R is the tube radius.

Using a mass balance with a first-order reaction,

$$vCdA = v\left(C + \frac{dC}{dL} \, dL \right) dA + k_r C dA dL \quad (6.114)$$

and

$$v\frac{dC}{dL} + k_r C = 0 \quad (6.115)$$

Also,

$$2\overline{V} \left[1 - \left(\frac{r}{R} \right)^2 \right] \frac{dC}{dL} + k_r C = 0 \quad (6.116)$$

and

$$\ln\left(\frac{C_0}{C} \right) = \frac{k_r L}{2\overline{V} \left[1 - \left(\frac{r}{R} \right)^2 \right]} \quad (6.117)$$

Solutions of eq 6.117 for a first-order reaction are presented in Table 6.4. Because the change in values in the left column are due solely to a change in L, we can see the effect of reactor length: the longer the reactor (the greater L is), the more severe the effect of the velocity profile.

TABLE 6.4. Effect of Plug and Laminar Flow on the Exit Concentrations for a First-Order Reaction

$\dfrac{k_r L}{2\overline{V}}$	Laminar Flow[a]	Plug Flow[a]
0.01	0.9810	0.9802
0.10	0.8328	0.8187
0.50	0.4432	0.3679
1.50	0.2194	0.1353
2.00	0.0603	0.0183

[a]The results shown are solutions for eq 6.117.

SOURCE: Reproduced with permission from reference 4. Copyright 1957 AIChE.

NONISOTHERMAL CONDITIONS AND FLOW REACTORS

When designing nonisothermal flow reactors, we must combine the energy balance with the kinetic statement. The resulting solution, which is complex, is illustrated in the following example.

EXAMPLE 6.10. NONISOTHERMAL TUBULAR REACTOR VOLUME CALCULATIONS REQUIRE A TRIAL-AND-ERROR SOLUTION

A gas mixture at a feed rate of 9.1 kg-mol/h (40 mol% E; 40 mol% F; 20 mol% inert materials) is reacted in a tubular flow system with a 10-cm inside diameter at a pressure of 5.075×10^5 N/m^2 and an initial temperature of 833 K.

$$E + F \rightarrow G$$

The specific reaction rate k_r is given by

$$k_r = (2.46 \times 10^9) e^{-\frac{E}{RT}} \frac{m^3}{\text{kg-mol-s}}$$

where $E = 1.3787 \times 10^5$ kJ/kg-mol. Also

$$C_{P_E} = 7{,}743.8 \ \frac{J}{\text{kg-mol-K}}$$

$$C_{P_F} = 7{,}743.8 \ \frac{J}{\text{kg-mol-K}}$$

$$C_{P_{\text{inerts}}} = 6{,}453.1 \ \frac{J}{\text{kg-mol-K}}$$

$$C_{P_G} = 12{,}906.2 \ \frac{J}{\text{kg-mol-K}}$$

and

$$\Delta H_{reaction} = 5.34 \times 10^7 \ \frac{J}{\text{kg-mol}} \ \text{at 277 K}$$

For an overall heat transfer coefficient U of 8.76 W/m^2-K, and a T_m of 890 K, we must find the relation of reactor volume and temperature to conversion.

The kinetic statement for the reaction rate is

$$-r_E = k_r \left(\frac{n_E}{V}\right)\left(\frac{n_F}{V}\right)$$

but n_E equals n_F so that

$$-r_E = k_r \left(\frac{n_E}{V}\right)^2$$

$$-r_E = k_r \frac{[n_{E_0}(1 - f_E)]^2}{V^2}$$

Because the gas mixture is in the ideal gas range,

$$V = \frac{RT n_{total}}{P} = \frac{RT}{P} (n_{t_0} + n_{E_0} f_E \gamma)$$

and

$$-r_E = \frac{P^2 k_r [n_{E_0}(1 - f_E)]^2}{(RT)^2 (n_{t_0} + n_{E_0} f_E \gamma)^2}$$

The enthalpy balance is

$$-[n_{E_0} C_{P_E} + n_{F_0} C_{P_F} + n_{\text{inert materials}} C_{P \text{inert materials}}](833 - 277) + (\Delta H_{reaction})(n_{E_0} f_E)$$
$$+ [2 n_{E_0}(1 - f_E) C_{P_E} + n_{E_0} f_E C_{P_G} + n_{\text{inert materials}} C_{P \text{inert materials}}] \ (T - 277)$$
$$- \frac{4U}{D} \int_0^{f_E} \frac{(890 - T) df_E}{-r_E} = 0$$

The last term results because the heat transfer per mole of feed is

$$Q = \int \frac{U(dA)(T_m - T)}{-r_E dV_r} \ df_E$$

and because surface area is $(\pi D)dL$ and dV_r is $(\pi D^2/4)dL$.

Then taking 1 kg-mol of feed and substituting appropriate terms,

$$T = 277 \ \text{K} + \frac{4.16 \times 10^6 - 5.34 \times 10^7 \ n_{E_0} f_E + 350.4 \int_0^{f_E} \frac{(890 - T) df_E}{-r_E}}{7485.6 - 1032.56 f_E}$$

Also, from eq 6.94,

$$V_r = W \int_0^{f_E} \frac{df_E}{-r_E}$$

and

$$V_r = W \int_0^{f_E} \frac{(RT)^2 \, (n_{t_0} + \gamma n_{E_0} f_E)^2 \, df_E}{P^2 k_r [n_{E_0}(1 - f_E)]^2}$$

The preceding expressions are solved in the following manner:

1. Assume an incremental value of f_E and T_1.
2. Check the assumed T_1 value by substitution in the temperature equation. If T_1 is correct, both sides of the equation will be equal.
3. If T_1 does not check out for the f_E, repeat steps 1 and 2.
4. After you have developed a tabulation of $(f_E - T)$ data, calculate V_r as a function of f_E because T is known at each f_E).

Calculated curves for f_E and T as well as V_r and f_E are shown in Figures 6.12 and 6.13.

CONTINUOUS-FLOW STIRRED TANK OR STEADY-STATE REACTORS

The continuous-flow stirred tank reactor (CSTR) has equal flow rates into and out of the system; it is in a steady state. A single reactor or a series of such reactors have special advantages that include flexibility, suitability for slow reactions, and

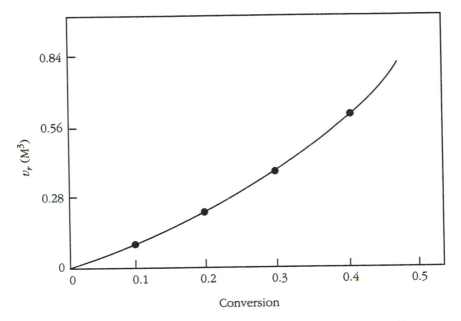

FIGURE 6.12. Calculations of required volume for a nonisothermal reaction in a flow reactor involve solutions of an energy balance and a kinetic statement. The resulting equations are complex and are solved by trial and error. The curve shown is the reaction volume required for a conversion in a flow reactor, based on conditions given in example 6.10. (Adapted from reference 10.)

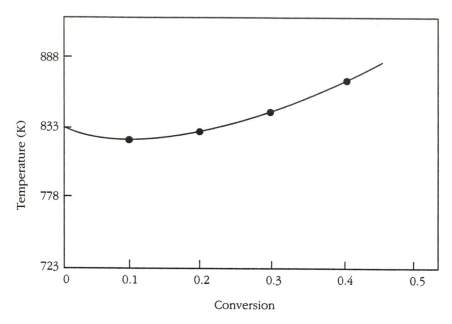

FIGURE 6.13. Reactor temperature vs. conversion for a flow reactor and nonisothermal conditions, as shown in Example 6.10. (Adapted from reference 10.)

the fact that they are an excellent compromise between batch and tubular flow reactors.

The basic approach in designing CSTRs is the overall mass balance

$$\text{input rate} = \text{output rate} + \text{accumulation} \tag{6.118}$$

For a battery of n reactors (Figure 6.14), the mass balance on the nth reactor is

$$F_{n-1}C_{n-1} = F_n C_n + r_n V_n + \frac{dC_n V_n}{dt} \tag{6.119}$$

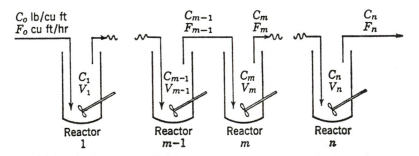

FIGURE 6.14. A battery of continuous-flow stirred tanks reactors. (Reproduced with permission from reference 10. Copyright 1989 Stanley Walas.)

where F is the volumetric rate of flow, V is the reactor volume, C is the appropriate concentration, and r_n is the rate of reaction in the nth reactor. At steady state,

$$\frac{dC_n V_n}{dt} = 0 \qquad (6.120)$$

If there is uniform overflow (F is constant), then

$$r_n = -\frac{F}{V_n}C_n + \frac{F}{V_n}C_{n-1} \qquad (6.121)$$

Rewriting V_n and r_n in terms of a residence time θ_n,

$$\theta_n = \frac{V_n}{F} \qquad (6.122)$$

and, substituting θ_n in eq 6.121 results in

$$r_n = -\frac{C_n}{\theta_n} + \frac{C_{n-1}}{\theta_n} \qquad (6.123)$$

Equation 6.123, together with an appropriate rate equation for a given reactor as a function of C_n,

$$r_n = k_r \phi(C_n) \qquad (6.124)$$

can be used to solve for the concentration change in the reactor.

If, for any stage, we can define the rate as a function of concentration for one reactant,

$$r_n = k_r \phi(C) \qquad (6.125)$$

then the number of stages needed to carry out a particular reaction can be found graphically (Figure 6.15).

The graphical solution involves first developing a curve for r (i.e., $-dC/dt$) vs. C. Next, we use a constant slope ($-F_0/V$ or $-1/\theta$) from each C, starting with C_0. Figure 6.15 shows that four reactors would be required to go from C_0 to C_4.

Figure 6.16 shows the effect of operating over a range of temperatures in the reactors. Residence time, θ, remains unchanged for each reactor.

The stirred tank system also lends itself to an analytical solution. For a first-order reaction, substituting ($r = k_r C$) in eq 6.123,

$$C_n + r_n \theta_n = C_{n-1} \qquad (6.126)$$

yields

$$C_n + k_r \theta_n C_n = C_{n-1} \qquad (6.127)$$

$$C_n = \frac{C_{n-1}}{1 + k_r \theta_n} \qquad (6.128)$$

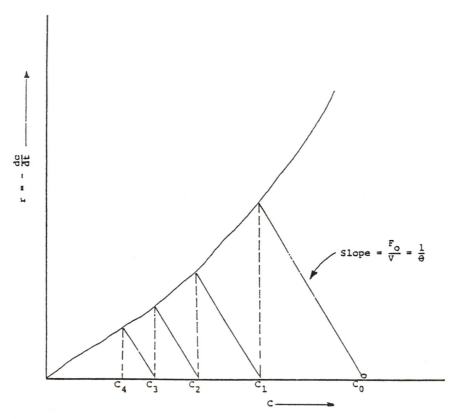

FIGURE 6.15. Graphical solution for the number of reactors needed in series for a similar reaction rate. First the curve is developed for r vs. C. Then the slope $-1/\theta$ is drawn from each concentration, starting with C_0. The result is that four reactors are needed. (Adapted from reference 10.)

or

$$C_1 = \frac{C_0}{1 + k_r\theta_1} \tag{6.129}$$

and

$$C_2 = \frac{C_1}{1 + k_r\theta_2} = \frac{C_0}{(1 + k_r\theta_1)(1 + k_r\theta_2)} \tag{6.130}$$

and so on. When all residence times are equal,

$$C_n = \frac{C_0}{(1 + k_r\theta)^n} \tag{6.131}$$

For a second-order reaction, substitution ($r = k_r C^2$) in eq 6.123,

$$C_n + k_r\theta_n C_n^2 = C_{n-1} \tag{6.132}$$

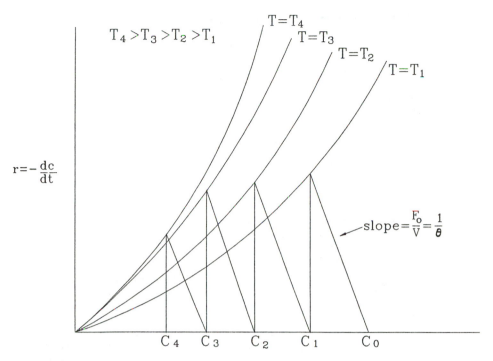

FIGURE 6.16. Graphical solution for reactors in series when the reaction in each reactor varies. Residence time remains the same, but the temperature varies.

Solving,

$$C_n = \frac{-1 \pm \sqrt{1 + 4k_r\theta_n C_{n-1}}}{2k_r\theta_n} \qquad (6.133)$$

Then, for example,

$$C_1 = \frac{-1 \pm \sqrt{1 + 4k_r\theta_1 C_0}}{2k_r\theta_1} \qquad (6.134)$$

and

$$C_2 = \frac{-1 \pm \sqrt{1 + 4k_r\theta_2 C_1}}{2k_r\theta_2} \qquad (6.135)$$

Equation 6.134 for C_1 must be substituted for C_1 in eq 6.135 to put it in terms of C_0.

   Various graphical solutions for stirred-tank reactor systems are available. A set of graphs showing solutions for a second-order reaction are shown in Figure 6.17.

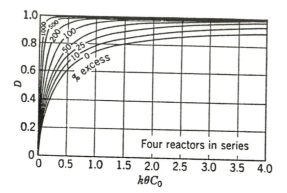

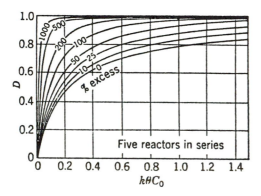

FIGURE 6.17. Graphical solutions for tank batteries already exist, as shown by these solutions for a second-order reaction, $r = k_r(C_A + E)$, where E is the excess of reactant B over A. (Reproduced with permission from reference 5. Copyright 1950 AIChE.)

EXAMPLE 6.11. CALCULATING CONCENTRATIONS AND REACTOR VOLUMES FOR A REVERSIBLE REACTION VIA A GRAPHICAL METHOD

An isothermal reaction is carried out in a steady-state reactor system with a flow rate of 7.87×10^{-4} m^3/s. The reaction is

$$2A \underset{k_2}{\overset{k_1}{\rightleftharpoons}} C + D$$

The initial concentration of A is 24.07 kg-mol/m^3 and those of C and D are both zero. The k_1 is 1.731×10^{-4} m^3/kg-mol-s and K_e is 16.

For 80% of equilibrium conversion, we must find (a) the vessel volume if only one is used and (b) the number of vessels of equal volume needed if volume is restricted to 1/10 of the volume of part a.

First, we must find the equilibrium concentration y_e of C and D,

$$K_e = \frac{(y_e)^2}{(24.07 - 2y_e)^2} = 16$$

$$y_e = 10.7 \text{ kg-mol/m}^3$$

The final desired concentration is $0.8y_e$ or 8.56 kg-mol/m^3. Thus

$$C_{A \text{ final}} = [24.07 - 2(8.56)] \text{ kg-mol/m}^3$$

$$C_{A \text{ final}} = 6.95 \text{ kg-mol/m}^3$$

The rate equation is

$$-\frac{dC_A}{dt} = k_1\left(C_A^2 - \frac{C_C C_D}{K_e}\right)$$

$$-\frac{dC_A}{dt} = \left(1.731 \times 10^{-4}\ \frac{\text{m}^3}{\text{kg-mol-s}}\right)\left[(24.07 - 2y)^2 - \frac{y^2}{16}\right]\left(\frac{\text{kg-mol}}{\text{m}^3}\right)^2$$

It is then possible to plot $-dC_A/dt$ vs. corresponding C_A values ($C_A = 24.07 - 2y$) (Figure 6.18). By erecting a vertical line to the curve at $C_{A \text{ final}}$, we can locate the point to which $C_{A \text{ initial}}$ is joined. The resultant line $-1/\theta$ has a slope of -4.44×10^{-4} s^{-1}. Thus,

$$\text{slope} = -4.44 \times 10^{-4} \text{ s}^{-1} = -F/V_r = -1/\theta$$

$$V_r = F\text{ s}/4.44 \times 10^{-4} = 7.8 \times 10^{-4} \text{ m}^3/4.44 \times 10^{-4}$$

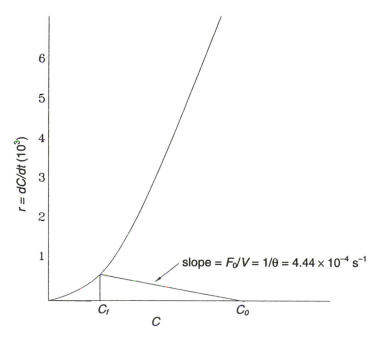

FIGURE 6.18. Graphical solution for Example 6.11, part a: one reactor to be used. Draw a vertical line from C_f to the $-dC/dt$ curve. The point so located is joined to C_0 to obtain the $-F/V$ or $-1/\theta$ line (slope $= -4.44 \times 10^{-4}$ s^{-1}).

and

$$V_r = 1.76 \text{ m}^3$$

This is the volume of one reactor.

Now if we use 1/10 the volume, then the volume of each reactor is 0.176 m³ and

$$-\frac{1}{\theta} = -\frac{F}{V_r} = -\frac{7.8 \times 10^{-4} \dfrac{\text{m}^3}{\text{s}}}{0.176 \text{ m}^3} = -\frac{4.44 \times 10^{-3}}{\text{s}}$$

We can start at $C_{\text{A initial}}$ and use the $-1/\theta$ values to find the C_A for the next stage (Figure 6.19). When completed (that is, $C_{\text{A final}}$ is reached), we find that four stages are required.

In spite of a tenfold volume change, the number of reactors required is not ten times the single stage of part a because reaction rate is not a linear parameter.

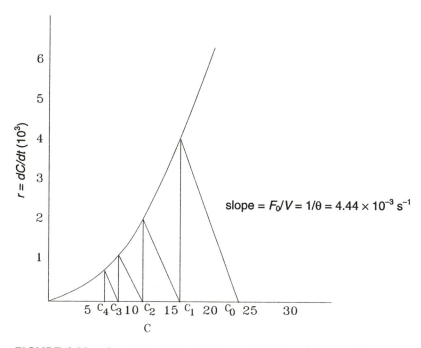

FIGURE 6.19. Graphical solution for Example 6.11, part b: several reactors to be used. Each reactor is to have 1/10 the volume in part a. We start at C_0 to obtain the $-F/V$ or $-1/\theta$ line (slope = -4.44×10^{-3} s⁻¹).

UNSTEADY-STATE OR SEMIBATCH REACTORS—ANALYTICAL AND ESTIMATION TECHNIQUES

As described in Chapter 3, the semibatch reactor is characterized by unsteady flows that affect the amount of conversion. The semibatch reactor is used for several things:

1. reacting systems that take place with large changes of heat; the semibatch method allows us to decrease the rate of exothermic reactions and to increase the rate of endothermic ones,
2. controlling an unfavorable by-product,
3. reactions affected by limited solubility, for example, a gas of limited solubility in a liquid; in such cases we can slowly add the limited solubility reactants and not inhibit the reaction,
4. systems in which a product needs to be removed continuously, and
5. cases that use pseudo-first-order kinetics.

The basic equation governing the semibatch reactor is a modification of eqs 6.118 and 6.119, which results in a form similar to eq 6.123.

$$\frac{dC_n}{dt} + r_n = \frac{C_{n-1}}{\theta} - \frac{C_n}{\theta} \tag{6.136}$$

or, for a first-order reaction,

$$\frac{dC_n}{dt} + \frac{C_n}{\theta} + k_r C_n = \frac{C_{n-1}}{\theta} \tag{6.137}$$

A general solution for the approach of a first-order reaction to steady state is shown in Figure 6.20. The ordinate indicates the fraction of steady state attained for a given time parameter.

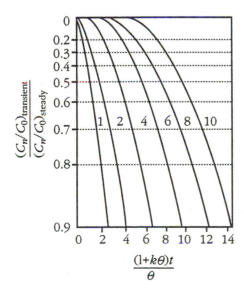

FIGURE 6.20. Unsteady-state behavior in stirred-tank reactors for a first-order reaction. (Reproduced from reference 6. Copyright 1951 American Chemical Society.)

EXAMPLE 6.12. CALCULATING THE REACTION COMPLETION OVER TIME REQUIRES ITERATIONS ON A COMPUTER

A reacting mixture of C and D in solution is placed in a stirred reactor of volume V_A. Substance D (concentration C_{D_0}) is then added to the tank at the rate of $W \, \text{m}^3/\text{s}$. The overflow is also at the rate of $W \, \text{m}^3/\text{s}$. If the kinetic statement is

$$C + D \rightarrow \text{products}$$

we must find the progress of the reaction.

We start with a mass balance across the system

$$\text{input} = \text{output} + \text{accumulation}$$

for each component.

For C the mass balance is

$$0 = WC_C dt + rV_A dt + V_A dC_C$$

For D

$$WC_{D_0} dt = WC_D dt + rV_A dt + V_A dC_D$$

Then,

$$(WC_{D_0} - WC_D - rV_A)dt = V_A dC_D$$

$$-(WC_C + rV_A)dt = V_A dC_C$$

and

$$\frac{dC_C}{dC_D} = \frac{-(WC_C + rV_A)}{WC_{D_0} - WC_D - rV_A}$$

This equation can be integrated, using the computer, to relate C_C and C_D. Then the C_C and C_D relationship can be used to relate concentrations to time.

GAS–SOLID REACTORS

The principal gas–solid reactors are the moving bed, the entrained bed, and the fluidized bed, all described in Table 6.5. Design of these reactors requires solution of equations that describe the kinetics and the transfer of energy and mass.

FIXED-BED REACTORS.   The fixed-bed reactor has been widely used in industry for such processes as ammonia synthesis, oxidation of sulfur dioxide, ethylene production, butane dehydrogenation, and reformation of naphthas. The fixed bed represents a case of combining flow, heat transfer, mass transfer, and chemical reaction. Because a packed bed is involved, the analysis of flow uses the superficial velocity in conjunction with the bed characteristics. We need to work with an energy and a mass

TABLE 6.5.   Reactors for Reactions Between Gases and Solids

Type	Description	Advantages
Fixed bed	Static bed of solids contacted by a flowing gas stream	Well-developed technology, useful for long residence times
Moving bed	Both solids and gases move	Can be used cocurrently or countercurrently, axial temperature gradient possible, allows constant catalyst regeneration
Entrained bed	Concurrent stream of gas and solids	Short residence times, excellent for very high temperatures
Fluidized bed	Solid particles suspended in a gas analogous to quicksand	Uniform temperature, excellent gas–solids contact, ease of solids handling

balance equation. Because such reactors are usually in towers, the energy balance in cylindrical coordinates is

$$C_P \overline{V} \rho \frac{\partial T}{\partial Z} + \Delta H_R r_C \rho_B = k_{\text{eff}} \left(\frac{1}{r} \frac{\partial T}{\partial r} + \frac{\partial^2 T}{\partial r^2} \right) \tag{6.138}$$

$\overline{V}$ is the gas superficial velocity, ρ is the gas density, ΔH_R is the heat of reaction, r_C is the rate of reaction, ρ_B is the bulk density, and k_{eff} is the effective bed thermal conductivity. The first term represents convective heat transfer in the axial (Z) direction, the second term is the thermal energy generated, and the last term is the heat conduction in the bed.

Similarly, for mass, the equation is

$$\overline{V} \frac{\partial \overline{V} C}{\partial Z} + \overline{V} r_C \rho_B = D_{\text{eff}} \left(\frac{1}{r} \frac{\partial \overline{V} C}{\partial r} + \frac{\partial^2 \overline{V} C}{\partial r^2} \right) \tag{6.139}$$

where D_{eff} is the effective diffusivity. The terms in eq 6.139 are, respectively, the convective mass transfer, the chemical reaction, and the diffusive mass transfer.

Solutions of equations 6.138 and 6.139 must be coupled and are complex. Solutions have been published for the oxidation of sulfur dioxide (*7*) and the synthesis of ammonia (*8, 9*).

FLUIDIZED-BED REACTORS.   The design of a fluidized bed is simpler than for a fixed bed because the fluidized bed rapidly attains a uniform temperature. Hence, only a mass balance is needed. Furthermore, the bed uniformity is so great that only longitudinal diffusion, diffusion along the z axis, is important. The governing equation (for a first-order reaction) is

$$\overline{V} \frac{\partial C}{\partial Z} + k_r C \rho_B = D_L \frac{\partial^2 C}{\partial Z^2} \tag{6.140}$$

where D_L is the effective diffusivity in the longitudinal or axial direction. The first term of eq 6.140 represents mass transport by convection in the bed, the second is for chemical reaction, and the third is for longitudinal diffusion.

Assuming an infinitely long reactor, we can solve eq 6.140.

$$Z = 0 \text{ at } C = C_0 \qquad (6.141)$$

and

$$Z = \infty \text{ at } C = 0 \qquad (6.142)$$

which gives

$$C = C_0 e^{\gamma Z} \qquad (6.143)$$

where

$$\gamma = \frac{1}{2} \frac{\overline{V}}{D_L} \left(1 - \frac{\sqrt{1 + 4 k_r \rho_B D_L}}{\overline{V}^2} \right) \qquad (6.144)$$

For a definite length and a concentration C_L,

$$C_L = C_0 e^{\gamma L} \qquad (6.145)$$

Also

$$X_L = \left(1 - \frac{C_L}{C_0} \right) = 1 - e^{\gamma L} \qquad (6.146)$$

The following example illustrates a specific case.

EXAMPLE 6.13. EVALUATING THE PERFORMANCE OF A FLUIDIZED BED

For a first-order fluidized reaction

$$A \xrightarrow{k} B$$

we must estimate the percent conversion for a 1.52-m-long reactor, given $k_1 = 0.0156$ m^3/s-kg catalyst, $\overline{V} = 0.3048$ m/s, and $\rho_B = 25.66$ kg/m^3. For a fluidized bed, an empirical relationship gives (in reciprocal feet)

$$\frac{\overline{V}}{D_L} = 2.6 \left(\frac{1}{\overline{V}} \right)^{0.6}$$

or

$$\frac{\overline{V}}{D_L} = 2.6 \text{ ft}^{-1} = 8.53 \text{ m}^{-1}$$

Then,

$$\frac{k_1\rho_B}{\overline{V}\left(\dfrac{\overline{V}}{D_L}\right)} = \frac{0.0156\ \dfrac{m^3}{s\text{-}kg_{catalyst}}\left(25.66\dfrac{kg}{m^3}\right)}{0.3048\dfrac{m}{s}\left(8.53\ \dfrac{1}{m}\right)}$$

$$\frac{k_1\rho_B}{\overline{V}\left(\dfrac{\overline{V}}{D_L}\right)} = 0.154$$

$$\gamma = \frac{1}{2}\left(8.53\ \frac{1}{m}\right)\left(1 - \sqrt{1 + 4(0.154)}\right)$$

$$\gamma = -1.15$$

$$X_L = 1 - e^{\gamma L}$$

and

$$X_L = 1 - e^{-1.15(1.52)}$$

Thus, the conversion is 82.6%. If longitudinal diffusion is neglected,

$$\frac{k_1\rho_B L}{\overline{V}} = 2.0$$

and

$$(X_L)_{ideal} = 1 - e^{-2} = 0.864$$

The result appears to be anomalous, but it is not because the longitudinal diffusion would reduce the reactant concentration in the bottom of the reactor and hence reduce chemical conversion.

The technique of Example 6.13 is not a design method but an evaluation technique: It compares conversions or lengths in a given system.

Design and scale-up of a fluidized bed is a complicated process that requires a number of steps. Failure to carry out this process carefully results in large-scale systems that fail to perform at required levels.

TYPES AND ADVANTAGES OF GAS–LIQUID REACTORS

A wide variety of reactors can be used for gas–liquid reaction systems, including the continuous-flow stirred tank reactor, batch reactor, plate column, and packed column (discussed in Chapters 5 and 6). Other appropriate devices are pipeline contactors, in which gas and liquid flow either co- or countercurrently, spray columns, aerators (which use turbine units), and trickle beds.

The trickle bed is a reactor that has found wide use in hydrotreating processes. In this reactor, liquid and gas streams flow downward cocurrently over the packed catalyst bed. The mechanisms in the trickle bed are complex. Particular factors of

TABLE 6.6.   Gas–Liquid Reactors

Type	Advantages
Continuous stirred tank	Excellent heat transfer Good for slow reactions Useful for hard-to-suspend slurries
Aerators	Best for slow reactions such as fermentation and for waste treatment plants
Spray columns	Good for cases that require low pressure drops Recommended if gas contains particles
Cocurrent pipeline	Excellent heat transfer Good temperature control Very good for irreversible reactions
Countercurrent pipeline reactor	Heat transfer is its most positive aspect
Packed columns	Excellent for corrosive systems Gives good service for foaming mixtures Has a low pressure drop
Plate columns	Holdup in column is advantageous for slow reactions Wide range of operating conditions
Trickle bed	For irreversible reactions and liquid-phase reactions Prevents unwanted side reactions

importance are flow distribution, diffusional resistances, and wetting characteristics.

Table 6.6 summarizes the optimum usages for the various gas–liquid reactors.

The reader should refer to the many excellent texts cited in the **Further Reading** section for additional information on the kinetics discussed, as well as for other topics not specifically considered.

REFERENCES

1. Walas, S. M. *Reaction Kinetics for Chemical Engineers;* McGraw-Hill: New York, 1959.
2. Smith, J. M. *Chemical Engineering Kinetics;* McGraw-Hill: New York, 1956.
3. Frost, A. A.; Pearson, R. G. *Kinetics and Mechanism;* John Wiley and Sons: New York, 1953.
4. Cleland, R.; Wilhelm, R. H. *AIChE J.* **1957**, *2*, 489.
5. Eldridge, J.; Piret, E. L. *Chem. Eng. Prog.* **1950**, *47*, 290.
6. Mason, L.; Piret, E. L. *Ind. Eng. Chem.* **1951**, *43*, 1210.
7. Baron, T. *Chem. Eng. Prog.* **1957**, *48*, 118.
8. Van Heerden, W. *Ind. Eng. Chem.* **1952**, *42*, 1242.
9. Adams, P.; Comings, L. W. *Chem. Eng. Prog.* **1953**, *49*, 359.
10. Walas, S. M. *Reaction Kinetics for Chemical Engineers;* Butterworth: Newton, MA, 1989.
11. Perry, R. H.; Pigford, R. L. *Ind. Eng. Chem.* **1953**, *45*(6), 1247.
12. *Chemische Technologie*, 5 vols.; Winnacker, Weingaertner, Eds.; Carl Hanser Verlag: Munich, Germany, 1950–1954.

FURTHER READING

Carberry, J. J. *Chemical and Catalytic Reactor;* McGraw-Hill: New York, 1976, out of print.

Danckwerts, P. V. *Gas–Liquid Reactions;* McGraw-Hill: New York, 1970, out of print.

Denbigh, K. G.; Turner, J. C. *Chemical Reactor Theory: An Introduction*, 3rd ed.; Cambridge University Press: New York, 1984.

Hill, C. G. *An Introduction to Chemical Engineering Kinetics and Reactor Design;* John Wiley and Sons: New York, 1976.

Hougen, O. A.; Watson, K. M. *Kinetics and Catalysis;* John Wiley and Sons: New York, 1946, out of print.

Levenspiel, O.; Kunii, D. *Fluidization Engineering;* reprint of 1969 ed.; Krieger, 1976.

Levenspiel, O. *Chemical Reaction Engineering*, 2nd ed.; John Wiley and Sons: New York, 1972.

Smith, J. M. *Chemical Engineering Kinetics*, 3rd ed.; McGraw-Hill: New York, 1980.

7

Process Control and Data Acquisition

As the chemical industry moved from a cottage enterprise to a large-scale endeavor, it needed operational stability and optimum productivity. Initially, these needs were met by increasing the number of operators who watched over the many variables of complex processes. This solution proved satisfactory through the 1930s.

However, economics, higher capacity plants, and severe operating conditions soon required new approaches to the industry's needs. These approaches were developed by introducing automatic control units to chemical and petroleum processing. Automatic controls reduced dependence on operators and provided the capability to deal with intricate processes.

Initially, control devices were placed in operation by applying experience or empirical techniques. As time went on, a scientific approach developed that led to innovative applications. This approach is known as *process dynamics* and its overall application as *systems engineering*.

Process control in chemical engineering is so important that a course in controls is now part of the undergraduate curriculum. Research in process dynamics and process control involves sizable numbers of faculty.

CONTROL APPLICATIONS AND TYPES OF SYSTEMS

Control is the organization of a function for a specific purpose. Control systems are widespread in chemical and petroleum process plants. All control systems change with time. These systems are information processing devices receiving information, acting on it, and generating an output. All such systems are integrated, requiring a systems approach.

Control systems in chemical process industries meet three main needs: (1) minimizing and regulating the influence of external disturbances, (2) guaranteeing a stable process, and (3) optimizing process performance.

MINIMIZING AND REGULATING EXTERNAL DISTURBANCE

When a change in flow rate, pressure, or temperature alters the behavior of a process unit, a control system suppresses the disturbances.

CREATING STABILITY

Most processes are stable and self-regulating; Figure 7.1 is the schematic of a stable system's response to some variable x with time. When the output of a process grows larger, it is considered unstable. A highly exothermic chemical reaction is a system that could be unstable. Unstable processes, in contrast to stable ones, exhibit the type of behavior shown in A, B, or C of Figure 7.2.

OPTIMIZING PRODUCTION OF PRODUCT

Optimization of productivity has an economic impact. Well-designed control systems allow the process variables to be changed logically to optimize the process and enhance the economic results.

TYPES OF CONTROL

Control systems are divided into two categories: *servomechanisms* and *regulators.* Servomechanisms change an output variable; regulators hold an output variable constant even if the input variable changes. Typical servomechanisms are those used for missile guidance or steering. Most chemical or petroleum process control systems are regulators. A chemical application that resembles a servomechanism is a program-controlled batch process using a preset pattern.

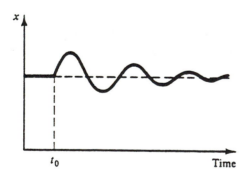

FIGURE 7.1. A stable system will show decreasingly smaller output to a given variable x over time. (Reproduced with permission from reference 1. Copyright 1984 Prentice-Hall.)

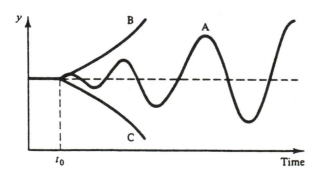

FIGURE 7.2. An unstable system's response to a variable increases over time. (Reproduced with permission from reference 1. Copyright 1984 Prentice-Hall.)

BASICS OF CONTROL SYSTEMS

Every control system has a variable that we wish to control, called a *controlled variable*, which we compare with a desired value, called a *reference value* or *set point*. Control is achieved by adjusting a third value, the *manipulated variable*.

FEEDBACK CONTROL

We can consider the simple control system shown in Figure 7.3, where the temperature of the exit stream from a heat exchanger is controlled by the flow of steam. This type of system uses feedback control.

The same heat exchanger control system is depicted simply with a block diagram in Figure 7.4. The feedback loop consists of the process, a measuring device, the comparator, the controller, and a final control element. In the loop shown, the following symbols apply: R is the reference or desired value, B is the measured value, E is the error $(R - B)$, M is the manipulated variable, and C is the controlled variable. Figures 7.3 and 7.4 depict the process and control system. The final control element is the steam-line control valve. The manipulated variables are the controller output signal M_1 and the valve output M_2. The controlled variable C is the exit stream temperature.

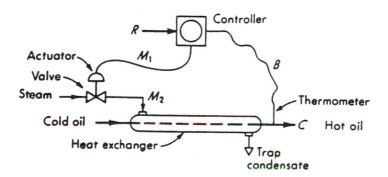

FIGURE 7.3. A control system can be depicted in a schematic. This heat exchanger has a regulator, which holds the temperature of the exchanger's output constant by controlling the steam input to the exchanger. (Reproduced with permission from reference 2. Copyright 1967 McGraw-Hill.)

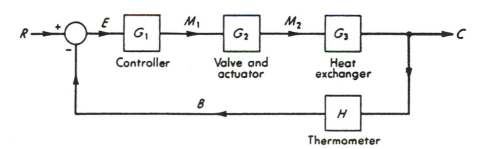

FIGURE 7.4. A block diagram is a simpler way to show how a control system operates. This is the block diagram for the heat exchanger control system shown in Figure 7.3. (Reproduced with permission from reference 2. Copyright 1967 McGraw-Hill.)

A controller mechanism consists of a comparator and the controller proper. The comparator examines the measured and desired variable values to determine the error. The controller changes the final control element's setting to reduce the error as rapidly as possible and to keep the system disturbance to a minimum. The control action to accomplish these purposes depends on the dynamics of the control loop. Control actions consist of some combination of three principal types: proportional, integral (or reset), and derivative (or differential).

TYPES OF CONTROL ACTIONS

Proportional control can be described as

$$M = M_0 + K_s e \tag{7.1}$$

in which M is the controller output, M_0 is the output with zero error, K_s is the sensitivity or gain, and e is the error.

A second mode of control is *integral* (or *reset*), in which

$$M = M_0 + K_I \int_0^t e\ dt \tag{7.2}$$

and K_I is a constant.

The third mode is *derivative* (or *differential*), which can be described by

$$M = M_0 + K_D \frac{de}{dt} \tag{7.3}$$

K_D is again a constant.

The three principal modes of control can be combined as indicated in Table 7.1; the behavior of combined control modes is illustrated in Figures 7.5–7.7. In Figure 7.5, we can see that proportional control continually shows a difference between the desired value (represented by zero on the y axis) and the control point (dotted line). The offset can be reduced by increasing K_s, although there is a limit because there will be oscillatory behavior. Proportional control must always have some offset.

Figures 7.6, 7.7, and Table 7.2 compare proportional (P), proportional plus dif-

TABLE 7.1.   Combined Control Modes

Mode	Symbol	Equation
Proportional plus integral	P + I	$M = M_0 + K_s e + K_I \int_0^t e\,dt$
Proportional plus differential	P + D	$M = M_0 + K_s e + K_D \dfrac{de}{dt}$
Proportional plus integral plus differential	P + I + D	$M = M_0 + K_s e + K_I \int_0^t e\,dt$ $+ K_D \dfrac{de}{dt}$

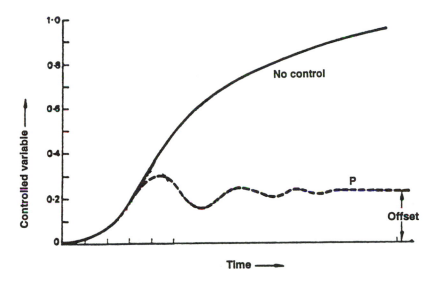

FIGURE 7.5.   Under proportional control, the controlled variable will always be offset by the desired value. In this graph, the desired value is zero, whereas the ultimate offset is 0.2. (Adapted from reference 3.)

ferential (P + D), and proportional plus integral plus differential (P + I + D) control. Each of the combinations gives the advantages shown. All of the combinations reduce the offset. For the P + D case, a higher gain is attained before instability. The net result is to lessen the offset for the P + D controller. The P + I device has an additional effect due to the integral model, which reduces the offset to zero. A P + I + D controller gives an even faster zero offset and also overcomes the P + I mode's disadvantages.

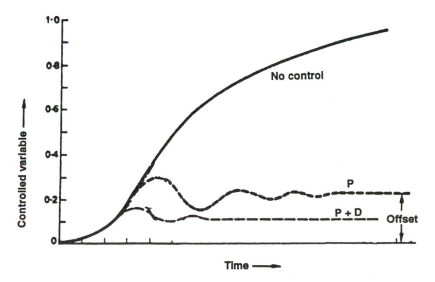

FIGURE 7.6.   Proportional plus derivative control reduces the amount of offset compared with proportional control. (Adapted from reference 3.)

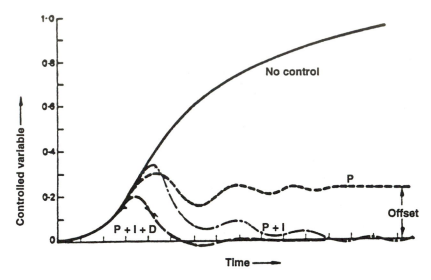

FIGURE 7.7. A comparison of different control combinations indicates that adding an integral to proportional or proportional and derivative control reduces or eliminates the offset, although oscillation will continue. (Adapted from reference 3.)

FEEDFORWARD CONTROL

Feedforward control measures the disturbances entering the process and adjusts the manipulated variables to control the process output. Feedback and feedforward control systems are compared schematically in Figure 7.8.

Whereas it is not as widespread as feedback control, feedforward control is valuable when applied to processes such as continuous-flow stirred tank reactor systems and distillation columns.

TABLE 7.2. Comparison of Control Modes

Mode	Advantages	Disadvantages
P	Simplicity	Offset
P + I	No offset	High maximum deviation
		Long periods of oscillation
P + D	Least oscillation	Offset
	Least maximum deviation	
P + I + D	No offset	Higher maximum deviation than P + D
	Second lowest oscillation and maximum deviation	Longer periods of oscillation than P + D
	Useful compromise	

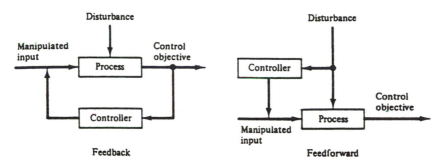

FIGURE 7.8. Comparison of feedback and feedforward control systems. (Reproduced with permission from reference 1. Copyright 1984 Prentice-Hall.)

PROCESS CONTROL EQUIPMENT

The two principal equipment items needed to implement process control are the *measuring element* and the final *control element*. Measuring elements are sensors that detect, to the best possible degree, the process variable selected for control. Final control elements are usually, but not always, automatic control valves, of which there are a number of varieties.

MEASURING ELEMENTS

Table 7.3 summarizes a list of measuring elements; many of the items are familiar to the practicing chemist. Other elements, such as the orifice or venturi meter, were described earlier in the chapter on fluid flow (Chapter 3). Equally familiar are two important terms related to measurement, namely, *accuracy* and *precision*. One additional term is *sensitivity*, which is how small a change in a process variable causes a response from an instrument. The largest change that can be made without eliciting a response is called the *dead zone*.

TABLE 7.3. Process Measuring Elements

Measuring Element	Principle Employed	Remarks
Flow		
Dahl flow tube	Pressure drop	
Kennison flow nozzle	Pressure drop	
Orifice flow meter	Pressure drop	
Venturi flow meter	Pressure drop	
Turbine	Turbine revolution	Very accurate
Hot-wire anemometer	Heat dissipation	High precision
Laser–Doppler velocimeter	Optics	Can measure velocity profiles
Ultrasound	Acoustics	

Continued

TABLE 7.3.   Process Measuring Elements—*Continued*

Measuring Element	Principle Employed	Remarks
Liquid level		
Liquid head pressure unit	Pressure	
Displacer unit	Position and displacement	
Float activated unit	Position and displacement	
Conductivity measurement	Electrical conductivity	
Electric behavior sensor	Electric behavior	
Sonic resonance	Acoustics	
Pressure		
Bellows device	Elastic deformation	
Bourdon tube	Elastic deformation	
Diaphragm	Elastic deformation	
Strain gauge	Elastic deformation	
Manometer	Fluid statics	
Piezoelectric and piezoresistivity device	Relationship between pressure and electricity	
Temperature		
Bimetallic thermometer	Expansion of metals	
Filled thermometer	Expansion of fluid	
Vapor pressure thermometer	Change of state of fluid	
Thermocouple	Electrical behavior	
Resistance thermometer	Electrical resistance	
Optical pyrometer	Monochromatic radiation	For temperatures >500 °C
Radiation pyrometer	Concentrated radiant energy	For temperatures >500 °C
Composition		
Emission spectroscope	Spectroscopy	
Ultraviolet analyzer	Spectroscopy	
Infrared analyzer	Spectroscopy	
X-ray absorption	Spectroscopy	
Mass spectrometer	Spectroscopy	
Microwave absorption spectrometer	Spectroscopy	Radio frequency range
Electron paramagnetic resonance spectrometer	Spectroscopy	Radio frequency range
Nuclear magnetic resonance spectrometer	Spectroscopy	Radio frequency range
Chromatographic analyzer	Adsorption	Long analysis times
pH meter	Chemical	
Oxidation–reduction potential	Chemical	
Absorption methods potential	Chemical	
Orsat analysis, etc.	Absorption	

TABLE 7.3.   Process Measuring Elements—*Continued*

Measuring Element	Principle Employed	Remarks
Composition		
Differential thermal analyzer, calorimeter	Thermal behavior	
Hydrostatic head instrument	Density	
Viscometer	Viscosity	
Wet bulb thermometer, hygrometer	Moisture content measurement	
Electrical test (all) and bridge circuits	Conductimetry	

CONTROL ELEMENTS

Measuring elements are connected to final control elements by transmission lines equipped with *transducers*. A transducer converts the measurement to a physical quantity such as a pneumatic or electrical signal. This signal is passed to the final control element. Transmission lines can be equipped with amplifiers to magnify the transmission.

Final control elements include relay switches, variable-speed compressors, variable-speed pumps, and automatic control valves.

The three main classifications of control valves are sliding-stem, pinch, and rotary-shaft valves. Valves with sliding stems include a variety of plug valves (Figure 7.9) and gate valves (Figure 7.10). The advantages and disadvantages of each of the valves are summarized in Table 7.4.

Pinch valves (Figure 7.11) are useful for systems that have slurries or corrosive fluids because they have no internal plug or gate that can be worn by the fluid.

Rotary shaft valves are of three types (Figure 7.12). The rotary plug valve gives a wide variety of flow characteristics and allows total shutoff. The butterfly valve, with either a rectangular or circular duct, uses the rotation of a vane to adjust flows from almost zero (vanes nearly perpendicular to flow) to almost maximum flow (vanes parallel to flow). The louver uses a set of rectangular vanes.

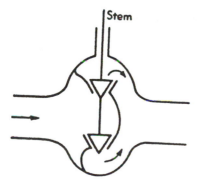

FIGURE 7.9.   A double-seated control valve is a type of plug valve because the seats, when lowered, reduce the flow. (Reproduced with permission from reference 2. Copyright 1967 McGraw-Hill.)

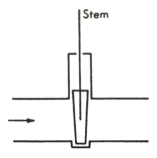

FIGURE 7.10. In a gate valve, the control system raises or lowers the gate to increase or decrease flow. (Reproduced with permission from reference 2. Copyright 1967 McGraw-Hill.)

SELECTION CRITERIA FOR CONTROL VALVES

Control valves are selected based on the *flow range* they can handle, the *flow-lift response*, and the *turndown*.

Rangeability is the ratio of maximum to minimum controllable flow through the valve. Control valves are not designed to close completely (the function of isolation is served by another type of valve) so that the rangeabilities vary. Sliding-stem control valves, for example, have a flow range on a scale of 20 to 70.

Lift is defined as the extent to which the valve is opened. The amount of flow depends both on the lift and the pressure drop across the valve. Figure 7.13 illustrates the flow-lift behavior of three different valves.

The turndown is the ratio of normal maximum flow to minimum controllable flow. Turndown values are about 70% of the rangeability. Control valves are selected by considering process rangeability, maximum specific range of flows, range of operating loads to be controlled, the fluid's character, and system pressure drops at minimum and maximum flow.

PROCESS DYNAMICS—DETERMINING CONTROL RESPONSE MATHEMATICALLY

The design of an optimum control system must take into account how all parts of a control loop will react to the load and to desired value variations. The reaction can be determined experimentally, but it can be found mathematically related to time.

TABLE 7.4. Advantages and Disadvantages of Sliding-Stem Valves

Valve	Advantages	Disadvantages
Single-seated plug	Can give zero flow	Requires large forces
Double-seated plug	Requires moderate force	
Equal percentage plug	Can give particular lift-flow characteristics	
Gate valve	Little change in flow direction	
	Minimizes settling and blockage	

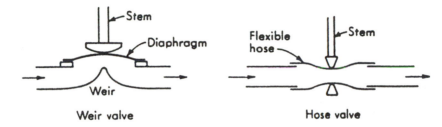

FIGURE 7.11. Because pinch valves have no internal gates or plugs, they work well in corrosive or slurried fluids. (Reproduced with permission from reference 2. Copyright 1967 McGraw-Hill.)

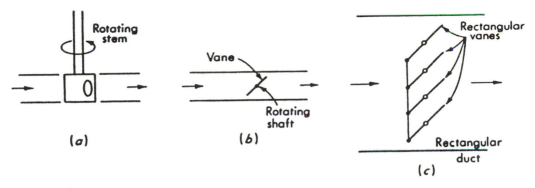

FIGURE 7.12. Rotary shaft valves include (a) plugs, (b) butterflies, and (c) louvers. (Reproduced with permission from reference 2. Copyright 1967 McGraw-Hill.)

LAPLACE TRANSFORMS

The dynamic behavior of the parts of the control loop can be complicated mathematically because many differential equations may be needed to describe the relationship between input and output. The approach used to simplify the calculation is a Laplace transform:

$$\bar{y} = \int_0^\infty y e^{-st} \, dt \tag{7.4}$$

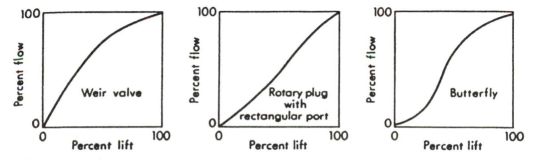

FIGURE 7.13. Every type of valve has a specific flow-lift behavior. Lift is the percentage of valve opening. (Reproduced with permission from reference 2. Copyright 1967 McGraw-Hill.)

where y is the variable that we relate to the independent variable t. The s is a parameter. The Laplace transform allows us to express differential equations algebraically.

An excellent example of simplification is the transfer function, which is defined as

$$\text{transfer function} = \frac{\text{Laplace transform of output}}{\text{Laplace transform of input}} \tag{7.5}$$

This equation gives a less complex input–output mathematical relationship. Furthermore, we can now write such a function for a given process.

The transfer function can be used to deal with either a first- or second-order system. An example of a first-order measuring device is a thermocouple, which measures temperature, or a device that measures the volumetric flow rate of a liquid flowing through a tank. The transfer function of a first-order system $G(p)$ is

$$G(p) = \frac{1}{1 + \tau p} \tag{7.6}$$

In eq 7.6, the transfer function represents the relationship between two quantities. For a thermocouple, the quantities are the temperatures of the surroundings and the thermocouple itself; for tank flow, the quantities are the input and output. The τ term is a time constant; for a thermocouple, τ is the product of the heat transfer resistance of the fluid film surrounding the thermocouple and the heat capacity of the device.

Second-order systems characteristically involve the relationship of motion and inertia. Examples of such systems are the distance–velocity lag in sampling or the behavior of a manometer. The basic transfer function for second-order systems is

$$G(p) = \frac{K'}{\tau^2 p^2 + 2\zeta\tau p + 1} \tag{7.7}$$

where K' is the gain of the system, and zeta (ζ) is the damping coefficient, damping factor, or damping ratio that characterizes the degree of damping of the transient responses.

Controllers also have transfer functions (Table 7.5).

TABLE 7.5. Laplace Transfer Functions for Controllers

Type	Transfer function
Proportional (P)	$G_c(p) = K_s$
Proportional plus integral (P + I)	$G_c(p) = K_s(1 + 1/\tau_I p)$
Proportional plus derivative (P + D)	$G_c(p) = K_s(1 + 1/\tau_D p)$
Proportional plus integral plus derivative (P + I + D)	$G_c(p) = K_s(1 + 1/\tau_I p + \tau_D p)$

FORCING FUNCTIONS—A METHOD OF CHECKING A FEEDBACK LOOP

Disturbances that can be applied externally, called forcing functions, are useful in the experimental and theoretical analyses of control systems. If forcing functions are applied without feedback, we have an *open-loop response*; if applied with feedback, we have a *closed-loop response*. An open-loop response involves the determination of an overall transfer function for the process by the control engineer. Forcing functions are important to the process control engineer. Some types of forcing functions are the step function (Figure 7.14), the sinusoidal function (Figure 7.15), and the pulse function (Figure 7.16).

The simplest and easiest forcing function to apply to a system is the step function. It is, however, a dangerous function because too large an input could put the system out of control. In treating the results, subtle behavior patterns may go undetected, causing significant errors at high frequencies for step testing.

Frequency testing involves using sinusoidal changes (Figure 7.15). This approach is time-consuming because it has to be performed at a number of frequencies. However, the advantages of frequency testing include minimal system disturbance and accurate measurement of a wide range of time constants.

The shape of the pulse in pulse testing can take many forms, including triangular and rectangular (Figure 7.16). This type of forcing function application requires

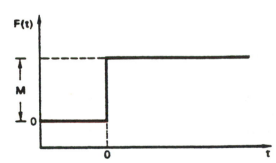

FIGURE 7.14. A step function is the simplest forcing function that can be externally applied to a control system. (Reproduced with permission from reference 3. Copyright 1979 Pergamon.)

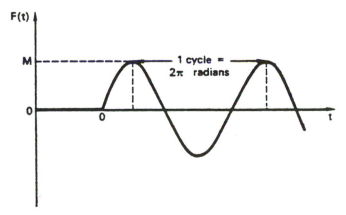

FIGURE 7.15. A sinusoidal function applied to a control system creates a minimum of disturbance. (Reproduced with permission from reference 3. Copyright 1979 Pergamon.)

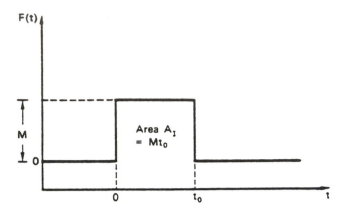

FIGURE 7.16. A pulse function can test a control system's response with one short input. Although this pulse is rectangular, a pulse can take on other forms, such as triangular. (Reproduced with permission from reference 3. Copyright 1979 Pergamon.)

excitation but not saturation of the system dynamics. Because a single pulse can excite all of the system frequencies, the testing can be completed with a single test. With pulse testing, highly sophisticated computational methods must be used.

PROCESS DYNAMICS DESIGN METHODS

A control engineer, typically a chemical engineer by training, is responsible for the analysis of processes as well as the analysis, behavior, and design of control systems. These tasks are accomplished by process dynamics, i.e., using frequency testing together with appropriate mathematical methods. Among these methods are the Routh–Hurwitz criterion, Bode plots, Nyquist plots, and the Cohen–Coon method. Implementation of process dynamics involves complex mathematical techniques that we shall describe to relate them to design.

All of the frequency testing methods deal with stability in control systems. We can consider an overall transfer function of a system. This function is

$$\text{overall transfer function} = \frac{G_A}{1 + G_B} \tag{7.8}$$

where G_A is the product of all transfer functions of all blocks in a system between the input and output. G_B is the product of all system transfer functions.

The characteristic equation of the system is

$$1 + G_B = 0 \tag{7.9}$$

which can be written as a polynomial

$$1 + G_B = 0 = C_0 p^n + C_1 p^{n-1} + C_{n-1} p + C_n \tag{7.10}$$

In eq 7.10, C_0 must be positive; if it is negative, the equation is multiplied by -1. At this point, the control engineer can apply one of the frequency testing methods.

ROUTH–HURWITZ CRITERION

To carry out the Routh–Hurwitz stability analysis (4,5) the first step is to check the coefficients $C_0 \ldots C_n$, which should all be positive. If any are negative, the system is unstable. When the coefficients are positive, we go to the second step, setting up a matrix:

Columns

Rows	1	2	3	4
1	C_0	C_2	C_4	C_6
2	C_1	C_3	C_5	C_7
3	A_1	A_2	A_3	
4	B_1	B_2		
5	C_1			
$n + 1$	Z_1	Z_2		

The capital letters are the following combinations:

$$A_1 = \frac{C_1 C_2 - C_0 C_3}{C_1} \qquad A_2 = \frac{C_1 C_4 - C_0 C_5}{C_1} \qquad A_3 = \frac{C_1 C_6 - C_0 C_7}{C_1}$$

$$B_1 = \frac{A_1 C_3 - C_1 A_2}{A_1} \qquad B_2 = \frac{A_1 C_5 - C_1 A_3}{A_1}$$

$$C_1 = \frac{B_1 A_2 - A_1 B_2}{B_1}$$

All terms in column 1 (C_0, C_1, A_1, B_1, etc.) must be positive for system stability.

Although it is useful, the Routh–Hurwitz criterion has its drawbacks: The transfer function must be known, and the degree of stability may not be possible to find. One of the other stability tests that uses frequency response methods is, therefore, preferable.

BODE PLOT

Bode plots, Nyquist plots, and the Cohen–Coon method all involve converting the transfer function by substituting iw for p so that a complex number $G(iw)$ with the following data results: (1) A magnitude $|G(iw)|$ that would be the magnitude ratio if the system is forced with a sine wave of frequency w and (2) a phase angle equal to the phase angle obtained with a forcing function of a sine wave with a frequency w.

A frequently used technique for stability analysis using the converted transfer function is the Bode plot or diagram (6,7). This plot consists of two diagrams. One diagram plots the phase angle versus w. The second diagram plots one of a number of possible ratios or moduli versus frequency (Table 7.6).

For a first-order system, the amplitude ratio (AR) and phase angle are defined as

$$AR = \frac{1}{1 + \tau^2 w^2} \tag{7.11}$$

and

$$\text{phase angle } \phi = \tan^{-1}(-\tau w) \tag{7.12}$$

Plots of these functions are shown in Figure 7.17. In part b, the value of the frequency at a phase angle of $-180°$ is an important parameter called the crossover frequency (w_{co}). The AR value of the system at w_{co} is then an index of stability by the Bode stability criterion. A feedback control system is unstable if the AR value at the crossover frequency is greater than one.

NYQUIST PLOT

Another technique for evaluating stability is the Nyquist plot and criterion. This plot has the imaginary part of the function $G(iw)$ as the ordinate and the real part of the same function as the abscissa. The Nyquist stability criterion states that a closed-loop response is unstable if the open-loop Nyquist plot encircles the point $(-1,0)$. Figure 7.18 gives examples of stable and unstable systems.

Frequency response data can also be used to establish controller settings. The Ziegler–Nichols method (9) uses data from the Bode plot to fix the ultimate gain and the ultimate period of the controller.

$$\text{ultimate period} = T_u = \frac{2\pi}{w_{co}} \tag{7.13}$$

and

$$\text{ultimate gain} = K_w = \frac{1}{\alpha} \tag{7.14}$$

TABLE 7.6.   Bode Plot Parameters

Parameter	Definition
Amplitude ratio	Effect of sine function on forced response amplitude
Magnitude ratio	Ratio of output amplitude to input amplitude
Modulus L	$L = \log$ modulus
	$L = 20 \log(\text{magnitude ratio})$

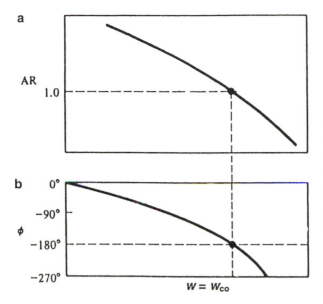

a

AR

1.0

b   0°

−90°

ϕ

−180°

−270°

$W = W_{co}$

FIGURE 7.17. When checking on the stability of a control system, information from Bode plots reduces the calculations. Two plots are needed: a, amplitude ratio vs. frequency and b, phase angle vs. frequency. If the amplitude ratio is greater than 1 at the crossover frequency w_{co}, the control system is unstable. (Reproduced with permission from reference 1. Copyright 1984 Prentice-Hall.)

where w_{co} is the crossover frequency and α is the overall amplitude ratio at w_{co}. The frequency and amplitude ratio are then used to get the terms for the control mode equation

$$M = M_0 + K_s + K_I \int_0^t edt + K_D \frac{de}{dt} \qquad (7.15)$$

written as

$$M = M_0 + K_s + \frac{1}{\tau_I} \frac{de}{dt} \qquad (7.16)$$

The terms for the controller setting arc shown in Table 7.7.

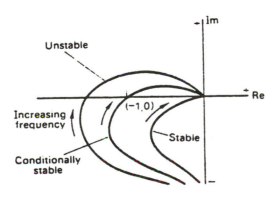

Unstable

Increasing frequency

(−1,0)

Stable

Conditionally stable

Im

Re

FIGURE 7.18. The Nyquist plot and criterion is another method for checking the stability of a control system. (Reproduced with permission from reference 3. Copyright 1979 Pergamon.)

TABLE 7.7. Ziegler–Nichols Controller Settings

Controller Type	K_s	τ_I	τ_D
Proportional (P)	$0.5\ K_w$		
Proportional + integral (P + I)	$0.45\ K_w$	$T_u/1.2$	
Proportional + integral + derivative (P + I + D)	$0.60\ K_w$	$T_u/2$	$T_u/8$

COHEN–COON METHOD

The Cohen–Coon method (*10*) can be used to find K_s, τ_s, and τ_D. This technique involves first using a response-time curve as shown in Figure 7.19. The times τ_{AD} (apparent system dead time) and τ_A (obtained from the tangent to the curve as shown), and u_s, the steady-state gain of the system, are combined to yield K_s, τ_I, and τ_D.

ADDITIONAL AND ADVANCED CONTROL METHODS

The preceding discussion was directed to feedback control systems. Other methods, however, merit discussion.

Feedforward control acts before the effect of a disturbance on a system, whereas feedback control acts in response to a disturbance. The advantages of feedforward control are its anticipatory action, its ability to control systems that are slow or have significant dead time, and the fact that it does not create instability in a closed-loop response. Its disadvantages include the facts that the process must be well understood, it cannot cope with unmeasured disturbances, it is insensitive to process parameter fluctuations, and it requires that all possible disturbances be characterized and measured. Table 7.8 gives various processes that use feedforward control. Also listed are disturbances and manipulated variables.

Another control system class has *multiple loops*. The loops may not be separate

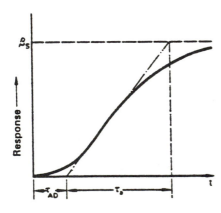

FIGURE 7.19. The Cohen–Coon method response vs. time curve can be used to find K_s, τ_s, and τ_D. τ_{AD} is the apparent dead time. (Reproduced with permission from reference 3. Copyright 1979 Pergamon.)

TABLE 7.8. Feedforward Controlled Processes

Process	Disturbances	Manipulated Variables
Heat exchanger	Flow rate and temperature of liquid entering tubes	Steam pressure for shell side
Drum boiler	Steam flow and feed water flow	Feed water flow
Distillation column	Feed flow rate and composition	Reboiler steam pressure and reflux
Continuous-flow stirred tank reactor	Inlet concentration and temperature	Product removal rate and coolant flow rate

and share either a measurement or a manipulated variable. One multiple-loop system is cascade control (Figure 7.20), in which the set point of a secondary (slave) controller is adjusted by the primary (master) controller. Cascade control systems are used with heat exchangers, furnaces, and distillation columns. The goal and benefit of cascade control systems is that disturbances occurring in the secondary loop can be corrected before the value of the primary controlled output changes.

Selective control systems involve one variable and several controlled outputs. A typical application prevents a process from exceeding a lower or upper limit; this is called override control. Another system using selective control is one in which the controller selects the highest value from among a set of similar measurements. This selective control system is known as auctioneering control. Selective control systems are effective in protecting boilers, compressors, and steam distribution systems.

Adaptive control is a self-optimization process in which the control system checks itself and compensates accordingly for disturbances. This type of control system involves extensive and complex computation by a digital computer. Adaptive control can be used for combustion systems, self-tuning regulators, and on-line tuning of proportional–integral controllers.

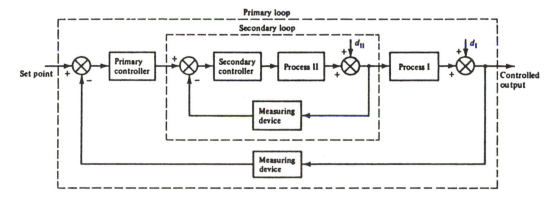

FIGURE 7.20.   A cascade control system has a secondary controller that receives input from a master or primary controller. (Reproduced with permission from reference 1. Copyright 1984 Prentice-Hall.)

If no direct measurement can be made of a process device's controlled output, *inferential control* is used. This approach requires a secondary measurement that infers, for example, a composition. (Incidentally, composition is a variable that is not easily determined because of the lack of effective measuring devices.) For this reason, inferential control is used most often with mass-transfer equipment or chemical reactors. Examples of its application are controlling product dryness without measuring it or controlling a distillation column without measuring either the feed or overhead composition.

The computer is widely employed in industrial process control. There are two basic types of computer control: direct digital control (D.D.C.) and supervisory digital control. In direct digital control, signals generated by a computer replace conventional controllers. In supervisory digital control, the computer only changes the set points or control parameter values of local controllers rather than directly replacing the controllers. Local controllers can be either of the analog or digital type.

If a computer is used in either direct or supervisory control, it must interact with an analog controller through an input/output (I/O) interface or a digital-to-analog converter (D/A converter). In supervisory control, these devices are not needed if the supervising computer communicates with another digital computer.

Direct digital control is particularly useful for cascade, feedforward, or adaptive control. Supervisory control finds wide application in optimizing processes.

DATA ACQUISITION BY COMPUTER

The computer has revolutionized the acquisition and storage of process data. Use of computers requires the following:

1. The process signal is transmitted to the computer without excessive signal noise.
2. The computers can accept the process signals and store the requisite data.
3. Data can be easily retrieved and read on an appropriate terminal.

In some cases, the computer can be directly interfaced to an analog signal source if the computer is not far from the signal source. If noise is a problem or if a large number of signals is to be handled, an intervening device must be used. In Figure 7.21, a schematic of a system involving a data-acquisition microcomputer is shown. Systems can be hardwired or the data can be transmitted by radio.

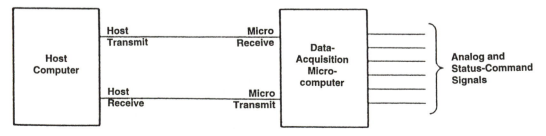

FIGURE 7.21. When distance is large between a controlling computer and the measuring device, a microcomputer can be used to gather the data and reduce signal noise. (Reproduced with permission from reference 11. Copyright Academic 1984.)

There are two basic types of digital signal transmission, parallel and serial signals. In parallel transmission, each bit of eight (a byte) goes to a separate conductor. Serial methods use a sequential method of transmission.

Computer usage involves languages (FORTRAN, BASIC, COBOL, etc.) for developing programs usable in a general manner with different computers. Computer languages can be subdivided into compiled and interpretive types. Compiled languages are those in which the machine code is created from the characters entered by the programmers. A typical compiled language is FORTRAN. Interpretive languages store programs as characters. Interpretation is done each time the program is used, which makes these languages seem slower than compiled languages. They do, however, have the advantages of simplicity. Interpretive BASIC is an example of this type of language.

Computers can operate as dedicated systems (having only one duty to perform) or time-sharing systems (having several clients). Time-sharing systems usually require the assignment of priorities. In many industrial situations, data acquisition and control may be accomplished with a computer network.

REFERENCES

1. Stephanopoulos, G. *Chemical Process Control;* Prentice-Hall: Englewood Cliffs, NJ; 1984.
2. Johnson, E. F. *Automatic Process Control;* McGraw-Hill: New York, 1967.
3. *Chemical Engineering,* Vol. 3; Richardson, J. F., Peacock, D. C., Eds.; Pergamon: Oxford, England, 1979.
4. Routh, E. J. *Dynamics of a System of Rigid Bodies,* 3rd ed.; MacMillan: London, England, 1977.
5. Hurwitz, A. *Math. Ann.* **1950,** *46,* 273.
6. Bode, H. W. *Bell Systems Tech. J.* **1940,** *19,* 421.
7. Bode, H. W. *Network Analysis and Feedback Amplifier Design;* Van Nostrand: New York, 1945.
8. Nyquist, H. *Bell Systems Tech. J.* **1932,** *11,* 126.
9. Ziegler, J. G.; Nichols, N. B. *Trans. ASME* **1942,** *64,* 759.
10. Cohen, G. H.; Coon, G. A. *Trans. ASME* **1953,** *75,* 827.
11. *Automated Stream Analysis for Process Control;* Manka, D. P., Ed.; Academic: Orlando, FL, 1984; Vol. 2, pp 63–91.

FURTHER READING

Buckley, P. S. *Techniques of Process Control;* reprint of 1964 ed.; 1979.
Considine, D. M. *Process Instruments and Control Handbook;* McGraw-Hill: New York, 1957, out of print.
Considine, D. M.; Ross, S. D. *Handbook of Applied Instrumentation;* reprint of 1964 ed.; Krieger, 1982.
Douglas, J. M. *Process Dynamics and Control;* Prentice-Hall: Englewood Cliffs, NJ, 1972, out of print.
Friedly, J. C. *Dynamics Behavior of Processes;* Prentice-Hall: Englewood Cliffs, NJ, 1972, out of print.
Harriott, P. *Process Control;* reprint of 1964 ed.; Krieger, 1983.
Harrison, T. J. *Handbook of Industrial Computer Control;* John Wiley and Sons: New York, 1972, out of print.

Minicomputers in Industrial Control; Harrison, T. J., Ed.; Instrument Society, 1978.

Johnson, E. F. *Automatic Process Control;* McGraw-Hill: New York, 1965, out of print.

Coughanowr, D. R.; Koppel, L. *Process Systems Analysis and Control;* McGraw-Hill: New York, 1965.

Kuo, B. C. *Analysis and Synthesis of Sampled Data Control Systems;* Prentice-Hall: Engelwood Cliffs, NJ, 1970, out of print.

Instrument Engineers Handbook, Vol. 1, Process Measurement; Liptak, B., Ed.; Chilton: Radnor, PA, 1982.

Luyben, W. L. *Process Modeling Simulation and Control,* 2nd ed.; McGraw-Hill: New York, 1972.

Murrill, P. *Automatic Control of Processes;* Intext Educational: Scranton, PA, 1967, out of print.

Ogata, K. *Modern Control Engineering;* Prentice-Hall: Englewood Cliffs, NJ, 1970.

Perone, S. P.; Jones, D. O. *Digital Computers in Scientific Instrumentation;* McGraw-Hill: New York, 1973, out of print.

Ray, W. H. *Advanced Process Control;* Butterworth: New York, 1989.

Computers in Manufacturing; Rembold, U. et al., Eds.; Marcel Dekker: New York, 1977.

Richardson, J. F.; Peacock, D. G. *Chemical Engineering;* Pergamon: Oxford, England, 1979; Vol. 3.

Schilling, G. D. *Process Dynamics and Control;* Holt, Rinehart and Winston: New York, 1963, out of print.

Shinskey, F. G. *Process Control Systems: Application, Design and Tuning;* McGraw-Hill: New York, 1989.

Smith, C. L. *Digital Computer Process Control;* Intext Educational: Scranton, PA, 1972, out of print.

Tou, J. T. *Digital and Sampled Data Control Systems;* McGraw-Hill: New York, 1959, out of print.

Weber, T. W. *An Introduction to Process Dynamics and Control;* reprint of 1973 ed.; Krieger, 1988.

Williams, T. J. *Systems Engineering for the Process Industries;* McGraw-Hill: New York, 1961, out of print.

8

Engineering Economics and Process Design

The practicing chemical engineer must be skilled in chemistry, physics, mathematics, and engineering science, and also in applied economics. The definition of chemical engineering by the American Institute of Chemical Engineers states that the profession requires "the application of the physical sciences together with economics . . . to fields that pertain directly to processes and process equipment."

Applied economics is considerably different than classical economics, which concerns itself with such topics as money supply and inflation. Applied economics emphasizes cost estimation, the time value of money, and the economic evaluation of processes. The economic evaluation of processes requires careful economic consideration of process alternatives. Mastery of applied economics is necessary for meaningful process decisions.

All chemical engineers have to be concerned with applied economics, although, depending on an engineer's responsibilities, he or she may approach economics differently. The differences are matters of scale rather than of content. A plant or process engineer, for example, concerned with pricing a piece of equipment may use the concept of economic alternatives. An engineer in a company's central engineering department may be concerned with the cost of an entire plant.

Chemists may not be required to use applied economics, but it will help them in formulating technical decisions.

TIME VALUE OF MONEY

The basis of proper economic decisions is a consistent system of the value of money. The value of a project can be evaluated on one of three cost bases: present worth, for which all future charges and costs are brought to the present and evaluated in today's dollars; annualized cost, for which costs or charges per year for the life of the project are examined; and future sum, for which all costs and charges are evaluated in terms of a future value. All of these valuation methods require understanding the time value of money because we must compare alternatives to each other and to the possibility of investing the money rather than selecting any project alternative.

INTEREST, THE COST OF MONEY

Application of the interest rate is necessary for any time value evaluation. The effect of compound interest on principal and the resulting future value is

$$S = P(1 + i)^n \tag{8.1}$$

where S is the future value of P, the principal invested at an interest rate i for n periods. The units of the interest rate and the number of periods n must match (per year, per week, or per day).

The time value of money is annualized by the following equation in which S is the future value of a series of uniform payments:

$$S = R(1 + i)^{n-1} + R(1 + i)^{n-2} + \ldots R(1 + i) + R \tag{8.2}$$

where R is a uniform payment for n periods. Equation 8.2 is abbreviated as

$$S = R\left[\frac{(1 + i)^n}{i}\right] \tag{8.3}$$

The bracketed term is the *series compound-amount factor*, which enables us to convert annual costs to future costs and vice versa.

Sometimes it is useful to know the present value of a stream of payments instead of the future value. The present value is

$$P = \frac{S}{(1 + i)^n} = R\left[\frac{(1 + i)^n - 1}{i(1 + i)^n}\right] \tag{8.4}$$

where P is the present value. The bracketed term is called the *series present-value factor.*

The reciprocal of eq 8.4 can tell us the annual payment R due if sum P is borrowed at interest rate i and is to be paid back in n periods.

$$R = P\left[\frac{i(1 + i)^n}{(1 + i)^n - 1}\right] \tag{8.5}$$

Here the bracketed terms is the *capital recovery factor.*

REPLACEMENT COST ESTIMATION

It is useful to estimate the cost of replacing a process unit at any given time period. This information is also needed to compare alternative equipment choices.

To compare alternative choices, we use the concept of replacing a process unit continually. We start by defining the present value P of the equipment as

$$P = S - C_r \tag{8.6}$$

where C_r is the replacement cost. By Combining eqs 8.1 and 8.6, we get

$$P = \frac{C_r}{(1 + i)^n - 1} \tag{8.7}$$

The capitalized cost K, the cost to replace the equipment, is

$$K = C_p + \frac{C_r}{(1 + i)^n - 1} \tag{8.8}$$

where C_p is the original equipment cost.

DEPRECIATION AND SALVAGE VALUE: DEPRECIATION METHODS AND SINKING FUNDS

All physical facilities undergo obsolescence and are subject to economic depreciation. Depreciation, the costs for using the facility, must be taken into account by the engineer. Depreciation is also a useful concept for chemists, particularly when they are involved with the operation of process units. To develop the depreciation formula, we need the facility's service life and the period of projected use. The rate of depreciation is also fixed by tax codes: A facility can be depreciated more slowly but never more quickly than tax laws permit.

Depreciation methods include straight line, declining balance, sum-of-years-digits, and sinking fund.

The *straight-line method* depreciates facilities at a constant rate. It is the value of the equipment divided by the service life

$$\text{depreciation} = \frac{\text{initial cost} - \text{salvage value}}{\text{service life}}$$

The *declining-balance method* depreciates facilities at an exponential ratio of the salvage value to the initial cost, using a declining balance factor X equal to

$$X = 1 - \left(\frac{\text{salvage value}}{\text{initial value}}\right)^{1/n}$$

where n is the life of the unit. At the end of the first year,

$$\text{value of unit} = \text{initial value} (1 - X)$$

At the end of the second year,

$$\text{value of unit} = \text{initial value} (1 - X)^2$$

In the final year,

$$\text{value of unit} = \text{salvage value} = \text{initial value} (1 - X)^n$$

The *sum-of-years-digits method* applies a depreciation factor,

$$\text{yearly depreciation factor} = \frac{\text{remaining useful years}}{\text{sum of total years}}$$

and

$$\text{yearly depreciation} = (\text{initial price} - \text{salvage value})(\text{yearly depreciation factor})$$

Both the sum-of-years-digits and the declining balance methods depreciate equipment more rapidly earlier in the equipment's service life than the straight-line method, although the total depreciation at the end of the equipment life is the same.

The *sinking-fund method* uses depreciation to accumulate money for the recovery of capital

$$\text{asset value after } c \text{ years} = V_i - (V_i - V_s)\left[\frac{(1 + i)^c - 1}{(1 + i)^n - 1}\right]$$

where V_i is the initial value, V_s is the salvage value, and n is the useful life.

ECONOMIC ALTERNATIVES

Engineers must reach decisions based on both technical feasibility and economic reality. This work requires evaluation of economic alternatives using the time value of money as shown by the following examples.

EXAMPLE 8.1. ESTIMATING EQUIPMENT REPLACEMENT COSTS

A decision is to be made as to which of two machines, A and B, is to be purchased. The initial purchase price and annual maintenance costs in dollars are

	A	B
Initial cost	20,000	15,000
Annual maintenance	4,000	5,000

We can assume a 10-year-life, no salvage value, and an annual interest rate of 12%. To compare alternatives, we annualize the costs using eq 8.5.

$$R = P\left[\frac{i(1 + i)^n}{(1 + i)^n - 1}\right]$$

Then,

$$R_A = 20,000\left[\frac{0.12(1.12)^{10}}{(1.12)^{10} - 1}\right] = 3,540$$

and

$$R_B = 15,000\left[\frac{0.12(1.12)^{10}}{(1.12)^{10} - 1}\right] = 2,655$$

The total annual cost of A = \$3,540 + \$4,000 = \$7,540. The total annual cost of B = \$2,655 + \$5,000 = \$7,655. Case A is preferable.

EXAMPLE 8.2. ESTIMATING THE PRESENT VALUE OF EQUIPMENT

Two process devices are to be compared. We can let X and Y designate the units.

	X(\$)	Y(\$)
Initial cost	15,000	11,000
Annual maintenance	1,500	2,000
Salvage value	5,300	4,900

We compare X and Y on the basis of present worth using a 15-year life and 10% interest. For the comparison, all the costs and the salvage value are brought to the present by applying

$$P = R\left[\frac{(1 + i)^n - 1}{i(1 + i)^n}\right]$$

for the annual maintenance costs.

	X	Y
Initial cost	15,000	11,000
Annual maintenance	$P = 1500\left[\frac{(1.10)^{15} - 1}{0.10(1.10)^{15}}\right]$	$P = 2000\left[\frac{(1.10)^{15} - 1}{0.10(1.10)^{15}}\right]$
	= 11,409	= 15,212
Subtotal	26,409	26,212
Salvage value	$P = \dfrac{5,300}{(1.10)^{15}} = 1,269$	$P = \dfrac{4,900}{(1.10)^{15}} = 1,173$
Total	25,140	25,039

Case Y is preferable because it has the lower present value.

GUIDELINES FOR COMPARING INVESTMENT ALTERNATIVES

When comparing alternative investments there are several rules that must be followed:

1. A preset minimum rate of return must be met by all of the alternatives.
2. A selected investment must return the minimum rate.

3. The investment of least cost that yields the desired rate of return should be accepted.
4. Alternatives must be compared to a base.
5. Total capital costs, not just fixed capital, must be used.
6. The minimum rate of return should be set at an interest rate if capitalized cost, annuities, etc., are used for comparison.

A technique frequently useful in comparing alternatives is incremental costs, which the following example illustrates.

EXAMPLE 8.3. COMPARING THE INCREMENTAL COSTS OF ALTERNATIVES

A company is considering adding a new unit that will require a $1,200,000 investment and yield an annual profit of $240,000. An alternative investment of $2,000,000 will yield $300,000. If the minimum rate of return is to be 14%, which is more favorable?

We can compare the rate of return of both investments to 14%,

$$\text{rate of return in case 1} = \frac{240{,}000}{1{,}200{,}000}(100) = 20\%$$

and

$$\text{rate of return in case 2} = \frac{300{,}000}{2{,}000{,}000}(100) = 15\%$$

Both seem to be acceptable. However, on an incremental basis,

$$\text{incremental rate of return} = \frac{(300{,}000 - 240{,}000)}{(2{,}000{,}000 - 1{,}200{,}000)}(100) = 7.5\%$$

Hence, investment 1 is clearly preferable.

EXAMPLE 8.4. COMPARING THE INCREMENTAL COSTS OF MULTIPLE UNITS

A process is to use an evaporation system that can be either single or multiple staged. The initial cost is $18,000 for each unit. Total operating costs are given below for each configuration.

Number of units	Cost/day ($)
1	319
2	203
3	176
4	170
5	173

The minimum acceptable investment return is 14%. How many units should be used?
Comparing two units to one,

$$\text{incremental rate of return} = \frac{(\$319 - 203)(300 \text{ days})}{(\$36,000 - 18,000)}(100) = 193\%$$

Because 193% far exceeds 14%, at least two units are preferable.
Comparing three units to two,

$$\text{incremental rate of return} = \frac{(\$203 - 176)(300 \text{ days})}{(\$54,000 - 36,000)}(100) = 45\%$$

This result indicates that three units are acceptable.
Comparing four units to three,

$$\text{incremental rate of return} = \frac{(\$176 - 170)(300 \text{ days})}{(\$72,000 - 54,000)}(100) = 10\%$$

Four units are unacceptable because the increment is less than 14%.
Next, we compare five units to three units:

$$\text{incremental rate of return} = \frac{(\$176 - 173)(300 \text{ days})}{(\$90,000 - 54,000)}(100) = 2.5\%$$

Three units are still preferred. Thus, the preferred alternative investment is three units.

EXAMPLE 8.5. COMPARING COMPLEX INVESTMENTS USING SEVERAL METHODS

We must compare three potential investments in a given project. A minimum annual return of 14% is required. We must use straight line depreciation.

	Service Life (years)	Initial Cost ($)	Working Capital Investment ($)	Salvage Value ($)	Annual Cash Flow to Project Alternatives ($)	Annual Expenses ($)
A	5	110,000	10,000	11,000	31,800	43,000
					32,800	
					36,800	
					40,800	
					45,800	
B	7	180,000	12,000	16,000	52,500	27,000
C	8	220,000	16,000	21,000	58,500	22,000

We can compare the alternatives on an annual rate of return basis. To calculate this rate, we subtract annual depreciation from annual cash flow. For example, for case B,

$$\text{annual return} = 52{,}500 - \left(\frac{\$180{,}000 - 16{,}000}{7 \text{ years}}\right) = \$29{,}072$$

Next, we compute the annual return for A.

$$\% \text{ average return on A} = \frac{\$17{,}600}{120{,}000}(100) = 14.7\%$$

The $17,600 was obtained by averaging the cash flow values for the five years of A, then subtracting annual depreciation. The $120,000 is the initial cost plus working capital.

$$\% \text{ incremental return on B} = \left(\frac{\$29{,}072 - 17{,}600}{\$192{,}000 - 120{,}000}\right)(100) = 15.9\%$$

and

$$\% \text{ incremental return on C} = \left(\frac{\$33{,}625 - 29{,}072}{\$236{,}000 - 192{,}000}\right)(100) = 10.3\%$$

Both A and B are acceptable.

Next, we compare all three alternatives on a present-worth basis. This comparison requires bringing the cash flow to the present and then subtracting the initial cost and working capital. For project A:

$$\text{net present worth for A} = \left(\frac{\$31{,}800}{1.14}\right) + \left(\frac{\$32{,}800}{(1.14)^2}\right) + \left(\frac{\$36{,}800}{(1.14)^3}\right)$$
$$+ \left(\frac{\$40{,}800}{(1.14)^4}\right) + \left(\frac{\$45{,}800}{(1.14)^5}\right) - \$120{,}000$$

net present worth for A = $5,840

For project B:

$$\text{net present worth for B} = 52{,}500 \left(\begin{array}{c} \dfrac{1}{1.14} + \dfrac{1}{(1.14)^2} + \dfrac{1}{(1.14)^3} + \dfrac{1}{(1.14)^4} \\[2mm] + \dfrac{1}{(1.14)^5} + \dfrac{1}{(1.14)^6} + \dfrac{1}{(1.14)^7} \end{array} \right)$$
$$- \$192{,}000$$

net present worth for B = $33,277

And for project C:

$$\text{net present worth for C} = 58{,}500 \left(\begin{array}{l} \dfrac{1}{1.14} + \dfrac{1}{(1.14)^2} + \dfrac{1}{(1.14)^3} + \dfrac{1}{(1.14)^4} \\[2mm] + \dfrac{1}{(1.14)^5} + \dfrac{1}{(1.14)^6} + \dfrac{1}{(1.14)^7} + \dfrac{1}{(1.14)^8} \end{array} \right)$$
$$- \$236{,}000$$

net present worth for C = \$35,499

On the basis of net present worth, C is the most favorable investment. Finally, we compare by capitalized costs.

$$\text{capitalized cost} = \frac{C_r(1 + i)^n}{(1 + i)^n - 1} + V_s + \frac{\text{annual expenses}}{i} + \text{working capital}$$

where C_r is the replacement cost (initial investment-salvage value) and V_s is the salvage value.

$$\text{capitalized cost for A} = \frac{(99{,}000)(1.14)^5}{(1.14)^5 - 1} + 11{,}000 + \frac{43{,}000}{0.14} + 10{,}000$$

capitalized cost for A = \$513,648

$$\text{capitalized cost for B} = \frac{(164{,}000)(1.14)^7}{(1.14)^7 - 1} + 16{,}000 + \frac{27{,}000}{0.14} + 12{,}000$$

capitalized cost for B = \$494,023

$$\text{capitalized cost for C} = \frac{(199{,}000)(1.14)^8}{(1.14)^8 - 1} + 21{,}000 + \frac{22{,}000}{0.14} + 16{,}000$$

capitalized cost for C = \$500,177

The preferred investment on the basis of capitalized cost is B.

Summarizing, we see that using a ranking system of 1, 2, or 3, with 1 being the best gives

Project	Annual Rate of Return	Net Present Worth	Capitalized Cost
A	2	3	3
B	1	2	1
C	3	1	2

The overall comparison shows that A is the least preferable and B is more favorable than C. Hence B is the project of choice.

The results of Example 8.5 clearly illustrate the importance of multiple comparisons in complicated alternatives. Only one comparison could cause an incorrect decision.

COST INDEXES—ESTIMATING PROCESS COSTS FROM EXISTING OPERATIONS

No costs remain stable. Pressures of inflation and deflation change them in tune with the general economy. As a result, in our estimates we must adjust any cost relative to some basis or index,

$$\text{present cost} = \text{initial cost}\left(\frac{\text{present index value}}{\text{index value at time of initial cost}}\right) \qquad (8.9)$$

A number of cost indexes exist (Table 8.1). Data for two of the indexes are given in Table 8.2 for the period 1973 to 1984. These years were selected because they represent a period of rapidly rising inflation followed by a leveling-off phase.

EXAMPLE 8.6. ESTIMATING THE COST OF A NEW UNIT BASED ON THE COST OF AN OLD ONE

A process unit cost $12,500 when installed in 1973. What would the cost have been in 1984?

Case 1 (Marshall–Swift):

$$1984 \text{ cost} = \$12,500\left(\frac{770.8}{344.1}\right) = \$28,000$$

Case 2 (Chemical Engineering plant cost):

$$1984 \text{ cost} = \$12,500\left(\frac{320.3}{144.1}\right) = \$27,784$$

Hence, the 1984 cost is about $28,000.

TABLE 8.I.   Cost Indexes

Index	Basis	Year of 100[a]	Where Published
Marshall–Swift	Installed equipment costs (all industry, process industry)	1926	*Chemical Engineering*
Engineering News-Record	Construction costs	1913, 1949, or 1967	*Engineering News-Record*
Nelson–Farrar Refinery Construction Index	Petroleum industry construction costs	1946	*Oil and Gas Journal*
Chemical Engineering plant cost index	Chemical plant construction costs	1957–1959	*Chemical Engineering*

[a]Year of 100 is an arbitrary number used to set the index (e.g., 1973 is 3.441 times 1926 on Marshall–Swift).

TABLE 8.2.   Data from Two Cost Indexes

Year	Marshall–Swift	Chemical Engineering Plant
1973	344.1	144.1
1974	398.4	165.4
1975	444.3	182.4
1976	472.1	192.1
1977	505.4	204.1
1978	545.3	218.8
1979	599.4	233.7
1980	659.6	261.2
1981	721.3	297.0
1982	745.6	314.0
1983	760.8	316.9
1984	770.8[a]	320.3[b]

[a]First quarter 1984.

[b]January 1984.

EQUIPMENT COST—CAPACITY AND SCALING FACTORS, SIX-TENTHS RULE

Process equipment costs are based on unit capacities, including factors such as heat transfer area in a heat exchanger, number of stages in a column, drying surface, and horsepower of a pump. Typical correlations between capacity and cost are given in Figures 8.1 and 8.2 for centrifuges used to separate inorganic and organic chemicals (*1*). Alloy construction markedly increases the cost for a given capacity.

SOURCES OF CAPACITY COST DATA

An article by R. S. Hall, J. Matley, and K. J. McNaughton (*1*) lists cost–capacity data for storage tanks, receivers, heat exchangers, tank vent condensers, reactors (dimple jacketed), reactor heating systems, blending tanks, centrifugal pumps, compressors and drivers, distillation columns, sieve trays, valve trays, and packed towers. Earlier cost–capacity sources can be found (*2, 3*) that contain comprehensive lists of data.

USING SCALING FACTORS TO ESTIMATE EQUIPMENT COSTS

When cost–capacity correlations are not available, we can estimate equipment costs by scaling factors:

$$\text{cost of unit 2} = \text{cost of unit 1}\left(\frac{\text{capacity of unit 2}}{\text{capacity of unit 1}}\right)^n \qquad (8.10)$$

where n is an exponent that can vary from 0.3 to 1.2. (Power factors for various

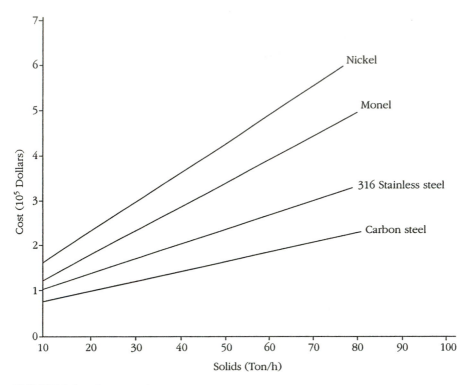

FIGURE 8.1. Cost as a function of capacity for centrifuges for inorganic service. These data are for solid bowl, screen bowl, and pusher types. (Adapted from reference 1.)

items are listed in Table 8.3.) If the exact exponent is not known, then a figure of 0.6 is used (i.e., the six-tenths rule).

The use of capacity indexes and charts is illustrated in Examples 8.7 and 8.8.

EXAMPLE 8.7. USING INDEXES TO ESTIMATE COSTS

In January 1979 the Ajax Corporation purchased a 100-ft^2 fixed tube sheet heat exchanger for $4950. What is the cost of a similar unit of 300-ft^2 in January 1984?

From Table 8.3, we find that the exponent for a fixed tube and sheet heat exchanger is 0.4, which we use to adjust the cost for greater capacity

$$1979 \text{ cost of } 300\text{-ft}^2 \text{ exchanger} = \$4950\left(\frac{300}{100}\right)^{0.4} = \$7682$$

Next, we compute the January 1984 cost.

$$\text{January 1984 cost} = \$7682\left(\frac{320.3}{233.7}\right) = \$10,528$$

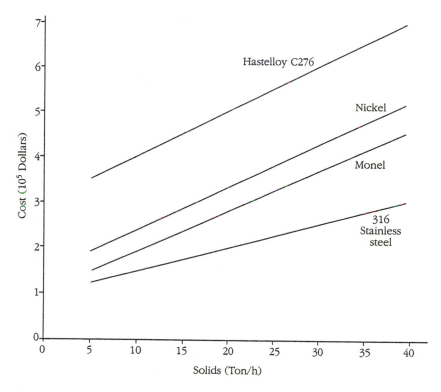

FIGURE 8.2. Cost as a function of capacity for centrifuges for organic service. These data are for solid bowl, screen bowl, and pusher types. (Adapted from reference 1.)

TABLE 8.3. Exponential Values for Equation 8.10 (Cost–Capacity Estimation)

Exponent	Process Unit Type
0.33	Kettles (250–800 gal.)
	Centrifugal pumps
	Reciprocating pumps
0.4	Vacuum crystallizers
	Atmospheric drum dryers
	Shell-and-tube fixed sheet vent exchanger
0.5	Double-cone rotary blenders
	Centrifugal fans (10^2–10^4 ft^3/min)
	Centrifugal separators
0.7	Explosion-proof motors (5–20 hp)
	Reciprocating compressors (100–400 ft^3/min)
0.8	Rotary compressors (100–1000 ft^3/min)
0.9	Sieve tray columns
1.0	Explosion-proof motors (20–200 hp)
1.2	Centrifugal fans (>20,000 ft^3/min)
	Bubble-cap tray columns

EXAMPLE 8.8. USING SCALING FACTORS TO ESTIMATE COSTS

We must estimate the cost of a 30-ton/h solid-bowl centrifuge for inorganic materials. The device is to be made of Hastelloy C276. Figures 8.1 and 8.2 give data for Hastelloy C276 only for organic systems. To estimate the cost for a centrifuge for inorganic materials, we use a ratio of the organic centrifuges. From Figure 8.2 (organic),

$$30 \text{ ton/h Hastelloy C276 centrifuge} = \$600,000$$

and

$$30 \text{ ton/h nickel centrifuge} = \$425,000$$

and from Figure 8.1 (inorganic),

$$30 \text{ ton/h nickel centrifuge} = \$300,000$$

Hence, the cost of a Hastelloy unit for inorganic materials is

Hastelloy 30 ton/h centrifuge for inorganic materials

$$= \$300,000 \left(\frac{600,000}{425,000} \right) = \$423,529$$

Published capacity charts (2, 3) can also be used. Update costs by using cost indexes. The six-tenths rule can also be used.

CAPITAL INVESTMENT ESTIMATING TECHNIQUES

The capital investment for a given process plant can be determined directly by totaling all direct (equipment, instrumentation, piping, and land) and indirect costs (supervision, overhead, contingency, and contractor's fees), but all such data must be available. In many instances, it is not possible to obtain a total directly. When data are not available, reliable estimates can be made using limited data and capital cost as a percentage of delivered equipment cost, multiplication factors using delivered equipment cost as a base, power factors with plant capacity ratio (that is, using eq 8.10), investment cost per unit capacity, and turnover ratios.

DATA FOR ESTIMATING COSTS BASED ON PERCENTAGES

Table 8.4 can be used to estimate plant costs based on the cost of delivered equipment. For example, assuming that we know the cost of the equipment, installation would cost approximately 45% of the equipment cost for a solid processing plant. It is possible to estimate overall capital investment by multiplying the total delivered equipment cost by 4.55, 4.87, or 5.69 (*see* the "Grand total" line of Table 8.4). This is the second estimation method listed above.

TABLE 8.4. Percentage Estimates of Plant Capital Costs

Costs	Solid Processing	Solid–Fluid Processing	Fluid Processing
Direct costs			
Equipment	100	100	100
Equipment installation	45	39	47
Instrumentation	9	13	18
Piping	16	31	66
Electrical	10	10	11
Buildings (with services)	25	29	18
Land improvements	13	10	10
Service facilities	40	55	70
Land	6	6	6
A. Total direct costs	264	293	346
Indirect costs			
Engineering, supervision	33	32	33
Construction expenses	39	34	41
B. Total direct and indirect costs	336	354	420
Fees, contingency, and working capital			
Contractor's fee (5% of B)	17	18	21
Contingency (10% of B)	34	36	42
C. Subtotal	387	413	483
Working capital (15% of C)	68	74	86
D. Grand total	455	487	569

ESTIMATING COSTS BASED ON MULTIPLES OF KNOWN COSTS

It is possible to use different multiplier factors for adding a facility that uses a given piece of equipment, for example: 4.0 for distillation columns, pressure vessels, pumps, and instruments; 3.5 for heat exchangers; 2.5 for compressors; and 2.0 for fired heaters (*4*).

USING POWER FACTORS FOR ESTIMATION

Table 8.5 lists power factors for plant capacity, which are used with eq 8.10, and dollars of fixed capital investment per ton of product (annualized) for various processes. The latter data are for the year 1978 and must be updated with an appropriate cost index.

TABLE 8.5. Capital Cost Data

Product or Process	Process Remarks	Fixed Capital Investment[a] per Annual Ton of Product	Power Factor for Plant Capacity Ratio
Chemical plants			
Acetic acid	CH_3OH and CO, catalytic	400	0.68
Acetone	Propylene–copper chloride catalyst	200	0.45
Ammonia	Steam reforming	150	0.53
Ammonium nitrate	Ammonia and nitric acid	30	0.65
Butanol	Propylene, CO, and H_2O, catalytic	500	0.40
Chlorine	Electrolysis of NaCl	340	0.45
Ethylene	Refinery gases	160	0.83
Ethylene oxide	Ethylene-catalytic	620	0.78
Formaldehyde (37%)	Methanol, catalytic	1000	0.55
Glycol	Ethylene and chlorine	1800	0.75
Hydrofluoric acid	Hydrogen fluoride and H_2O	500	0.68
Methanol	CO_2, natural gas, and steam	130	0.60
Nitric acid (high-strength)	Ammonia, catalytic	40	0.60
Phosphoric acid	Calcium phosphate and H_2SO_4	400	0.60
Polyethylene (high-density)	Ethylene, catalytic	2000	0.65
Propylene	Refinery gases	200	0.70
Sulfuric acid	Sulfur, catalytic	20	0.65
Urea	Ammonia and CO_2	80	0.70
Refinery units			
Alkylation (H_2SO_4)	Catalytic	1200	0.60
Coking (delayed)	Thermal	1600	0.38
Coking (fluid)	Thermal	1000	0.42
Cracking (fluid)	Catalytic	1000	0.70
Cracking	Thermal	300	0.70
Distillation (atm)	65% vaporized	200	0.90
Distillation (vac)	65% vaporized	120	0.70
Hydrotreating	Catalytic desulfurization	200	0.65
Reforming	Catalytic	1800	0.60
Polymerization	Catalytic	300	0.58

SOURCE: Data are from references 2 and 5–7.

[a]1978 dollars.

TURNOVER RATIO—ORDER OF MAGNITUDE ESTIMATES

The turnover ratio, gross annual sales divided by fixed capital investment, is a rapid method that gives an order of magnitude estimate for a plant. To use the ratio, we start with a prediction of annual sales and divide by the appropriate ratio (Table 8.6).

PRICING CHEMICAL PRODUCTS

An important economic task faced by chemical manufacturers is determining the price of a product. Gathering data to make a price determination is an integral part of market surveys. The pricing is usually done by the company's marketing division.

TABLE 8.6.   Turnover Ratios for Various Processes

Value	Process
0.2	Butadiene (butane)
	Butadiene (naphtha)
0.3	Allyl alcohol
	Butadiene (butylene)
	Sulfuric acid (anhydride)
0.4	Hexamethylenetetramine
	Synthetic ammonia
	Chlorine (nitrosyl chlorine)
	Soda ash
0.5	Butadiene (alcohol)
	Ethylene dichloride
	Alcohol (wood)
	Alcohol (sulfite liquor)
	Electrolytic caustic and chlorine
	Formaldehyde (hydrocarbons)
	Contact sulfuric acid (pyrites)
0.6	Neoprene
	Styrene
	Acetylene
0.7	Phenol, synthetic
	Furfural
	Alumina
0.8	Phthalic anhydride (*ortho*-xylene)
	Synthetic butanol
	Kraft paper
	Calcium cyanamide
	Contact sulfuric acid (smelter gas)
1.00	Benzaldehyde via chlorination
	Alcohol from grain
	Ethylene from refining gas

TABLE 8.6.   Continued

Value	Process
1.1	Methyl methacrylate resins
	Carbon black
1.22	Polyvinyl chloride (acetylene)
	Contact sulfuric acid
1.40	Synthetic glycerine
	Pentaerythritol
	Monoethylamine
	Phthalic anhydride
	Carbon disulfide
	Acetic anhydride (from acetic acid)
1.7	Oxalic acid
	Phosphoric acid (Dorr process)
	Disodium phosphate
1.9	Diphenylamine
	Hydrofluoric acid
	Alcohol (molasses)
	Sodium chloride (brine)
2.0	Toluene (hydroforming)
	Trichloroethylene
	Sodium silicate
3.0	Sodium bichromate
	Aluminum sulfate
4.5	Dialkyl phthalates
	Acrylonitrile (cyanohydrin)
	Methyl isobutyl ketone
	Formaldehyde (methanol)
8.3	Isopropyl alcohol
	Phenolic resin

It is, however, necessary for the practicing engineer to be able to estimate selling prices early in considering a process. This would also be useful for the chemist.

One method of determining price is to make an *exclusion chart*, which establishes the relationship between the production cost and the price of a chemical for a given period of time, such as a year. The result of such a survey for a given type of chemical is represented in Figure 8.3 (*8*). The shaded area in the figure is the exclusion zone, a region that should be avoided in relating price to volume. The chart also indicates that chemicals cannot be sold in volumes located above line AA unless their price is less than line AB. This information affords a useful method of estimating chemical selling prices.

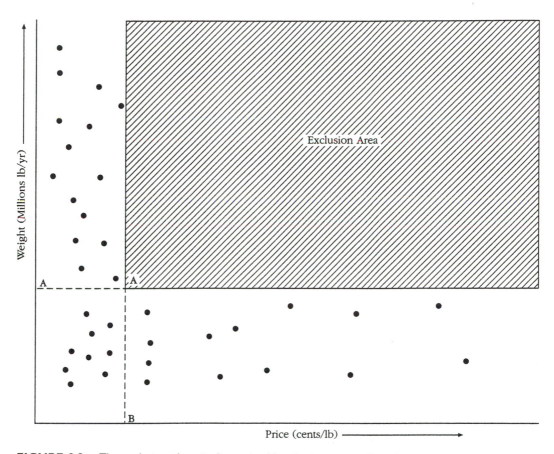

FIGURE 8.3.   The exclusion chart is determined by plotting cost and production data for a given chemical or class of chemicals, such as pesticides or engineering plastics. Data are usually available in trade literature. The exclusion area is the region where essentially no production cost data exist; this means that desired production for such costs is not economically sound. (Adapted from reference 8.)

PROCESS DESIGN

The actual initiation of a chemical process plant involves several stages which can be divided into research, feasibility studies, process development, process design, and construction and operation.

RESEARCH PHASE

Research may involve laboratory work or a literature search to find candidates for production or processes for producing certain chemicals. Chemists generally dominate at this step in the overall procedure.

FEASIBILITY STUDIES

After a potential candidate product or chemical process is identified, a feasibility study should be made. These studies serve as "go–no go" gauges. If the result of the study is negative, the project is canceled, and if positive, the project continues to the next stage. Information about raw materials needed, facilities available and needed, and possible profits must be considered.

The chemist is also involved with this step. Some examples of interaction are solving process problems such as controlling byproducts or preventing generation of hazardous and toxic wastes.

PROCESS DEVELOPMENT

During the process development stage, emphasis is on determining the basic technical data needed to design the process and its equipment. Both laboratory and pilot plant operations are required for this stage.

PROCESS DESIGN

When a process has been selected for further evaluation and the development aspects are completed, the design commences.

Initially, a *block diagram* is developed for the process. This diagram shows all the units (reactors, distillation columns, etc.) together with the flow path or route. Next, *material and energy balances* are made for the overall system and the individual units in the process. The results of this effort give the mass flow rates, temperatures, pressures, cooling and heating loads, steam requirements, and phase data.

Next, the specific units (distillation columns, extraction units, centrifuges, etc.) are designed to determine the appropriate sizes and capacities. Factors such as required stages, heat exchanger surface areas, and volumetric capacities are computed. The completion of this effort enables the designer to determine reliable costs of fixed equipment costs based on the capacities. Reliable costs lead to an excellent economic process evaluation.

Detailed physical designs of each unit then follow, referred to as "nuts and bolts" design. Completion of this effort allows a firm determination of equipment costs and in turn a final economic evaluation of the proposed plant.

CONSTRUCTION AND OPERATION

When all of the design has been completed and a favorable economic evaluation has been obtained, actual construction begins, followed by a start-up operation to bring the plant on stream.

A combination of in-house and outside talent build the plant. One approach is the *turn-key plant*, where a chemical or petroleum construction company carries out all stages of design from the feasibility study to the actual start-up for a client company. The other extreme is for the company itself to carry out the entire design, construction, and start-up with its own personnel; this approach is not as prevalent today as in the past. The most common approach is for a construction company to pick up the effort before detailed design but after development work has been completed.

Finally, the plant must go through start-up to bring it on line successfully. Both engineers and chemists are involved in this phase. After the start-up, optimization of the process begins to obtain the highest and most efficient level of operation. Again, chemists and engineers are involved in this effort.

CONCLUSION

This final chapter has added the important dimensions of applied economics and design procedures to the previously considered topics of chemical engineering thermodynamics, fluid flow, heat transfer, mass transfer, applied chemical kinetics, and process control and dynamics. All of this material represents the essence of the core of the undergraduate curriculum in chemical engineering (typically about 18 to 21 semester credit hours). As such, it should provide chemists, other scientists, and nonchemical engineers with a sound base in chemical engineering by emphasizing not only salient points, but also overall coherence to interrelate the subject areas. Again, as I pointed out in the preface, this text will help practicing professionals in the chemical and petroleum process industries to develop the chemical engineering skills needed to optimize their performance. Finally, it is hoped that the reader will go beyond the text to develop even greater skills in those aspects of chemical engineering covered in the treatment as well as more specialized areas.

REFERENCES

1. Hall, R. S.; Matley, J.; McNaughton, K. J. *Chem. Eng.* **1982,** *89*(7), 80.
2. Peters, M. S.; Timmerhaus, K. D. *Plant Design and Economics for Chemical Engineers;* McGraw-Hill: New York, 1980.
3. Chilton, C. H. *Cost Engineering in the Process Industries;* McGraw-Hill: New York, 1960, out of print.
4. Hand, W. E. *Pet. Refin.* **1958,** *37*, 331.
5. Guthrie, K. M. *Chem. Eng.* **1970,** *77*(13), 140.
6. Haselbarth, J. E. *Chem. Eng.* **1967,** *74*(25), 214.
7. Drayer, D. *Petro/Chem. Eng.* **1970,** *42*, 5, 10.
8. Zabel, H. W.; Marchitto, M. *Chem. Eng. Prog.* **1956,** *52*, 183.

FURTHER READING

Barish, N. N.; Kaplan, S. *Economic Analysis for Engineering and Managerial Decision Making;* McGraw-Hill: New York, 1978, out of print.

Bussey, L. E. *The Economic Analysis of Industrial Projects;* Prentice-Hall: Englewood Cliffs, NJ, 1978, out of print.

Engineering Economy, 8th ed.; DeGarmo, E. P. et al., Eds.; Macmillan: New York, 1989.

Grant, E. L.; Treson, W. G.; Leavenworth, R. S. *Principles of Engineering Economy*, 7th ed.; John Wiley and Sons: New York, 1982.

Jelen, F. C.; Black, J. *Cost and Optimization Engineering*, 2nd ed.; McGraw-Hill: New York, 1983.

Ostwald, P. F. *Cost Estimating;* Prentice-Hall: Englewood Cliffs, NJ, 1984.

Peters, M. S.; Timmerhaus, K. D. *Plant Design and Economics for Chemical Engineers*, 4th ed.; McGraw-Hill: New York, 1991.

Riggs, J. L. *Engineering Economics;* McGraw-Hill: New York, 1982.

Taylor, G. A. *Managerial and Engineering Economy;* D. Van Nostrand: Princeton, NJ, 1980, out of print.

Thuesen, G. J.; Fabrycky, W. J. *Engineering Economy*, 7th ed.; Prentice-Hall: Englewood Cliffs, NJ, 1989.

APPENDIX A

Dimensionless Groups and Scale Factors

One of the basic problems facing the engineer is scale-up. Without effective scale-up techniques, the time required for experimental process studies would be astronomical. The techniques used in modeling or scaling up use parameters called scale factors or dimensionless groups. These parameters represent ratios of factors or effects in a system.

For example,

1. Reynolds number = Re = $\dfrac{\rho V^2}{\mu V/D} = \dfrac{DV\rho}{\mu} = \dfrac{\text{inertial forces}}{\text{viscous forces}}$

where D is diameter, V is velocity, ρ is density, and μ is viscosity.

2. Froude number = Fr = $\dfrac{\rho V^2/D}{\rho g} = \dfrac{V^2}{gD} = \dfrac{\text{inertial forces}}{\text{gravity forces}}$

where g is the acceleration due to gravity.

3. $\dfrac{\text{Grashof number}}{(\text{Reynolds number})^2} = \dfrac{\text{Gr}}{(\text{Re})^2} = \dfrac{\rho \beta g(T_0 - T_1)}{\rho V^2/D} = \dfrac{\text{buoyancy forces}}{\text{inertial forces}}$

where β is the coefficient of thermal expansion.

4. (Prandtl number) (Reynolds number) = PrRe =

$\dfrac{\rho C_P V(T_0 - T_1)/D}{k(T_0 - T_1)/D^2} = \dfrac{\text{heat transport by convection}}{\text{heat transport by conduction}}$

where k is thermal conductivity, C_P is specific heat, and $(T_0 - T_1)$ is the temperature difference.

5. (Schmidt number) (Reynolds number) = ScRe =

$\dfrac{V(C_0 - C_1)/D}{D_{AB}(C_0 - C_1)/D^2} = \dfrac{\text{mass transfer by convection}}{\text{mass transfer by diffusion}}$

where D_{AB} is diffusivity and $(C_0 - C_1)$ is the concentration difference.

Furthermore, other dimensionless groups that are functions of the foregoing can be related to forces or effects.

$$\text{Nusselt number} = \text{Nu} = \frac{hD}{k} = \phi(\text{Re,Pr}) = \phi$$

$$\left(\frac{\text{inertial forces}}{\text{viscous forces}}, \frac{\text{heat transfer by convection}}{\text{heat transfer by conduction}} \right)$$

If the scale factors or dimensionless groups are the same for two different systems, then both systems are defined by identical dimensionless differential equations. If, in addition, the dimensionless initial and boundary conditions are the same (possible only if both systems are geometrically similar), then the two systems are mathematically identical and dynamically similar.

This means that it is possible in geometrically similar systems to use the scale factors or dimensionless group to scale up or scale down the given system.

The methods used to generate scale factors or dimensionless groups include the following:

1. intuition;
2. converting equations to dimensionless forms using characteristic lengths, masses, etc.;
3. the Buckingham pi theorem (*see*, for example, McAdams W. H. *Heat Transmission;* McGraw-Hill: New York, 1954 in Chapter 5.
4. force or effect ratios that require a knowledge of effects or forces to write appropriate ratios.

APPENDIX B

Physical Properties

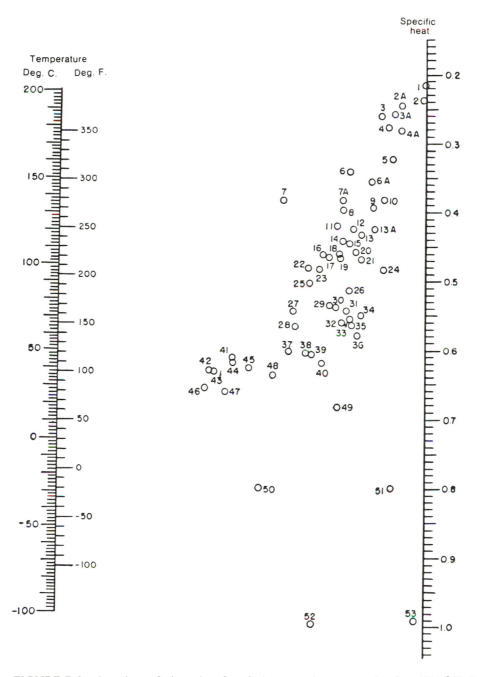

FIGURE B.I. Liquid specific heat data. Specific heats can be expressed as P.c.u./(lb)(°C), Btu/(lb)(°F), or calories/(g)(°C). (Reproduced with permission from *Heat Transmission;* McAdams, W. H., Ed.; McGraw-Hill: New York, 1954. Copyright 1954 McGraw-Hill.)

TABLE B.I. Key for Figure B.1

No.	Liquid	Temperature Range (°C) From	To
29	Acetic acid, 100%	0	80
32	Acetone	20	50
52	Ammonia	−70	50
26	Amyl acetate	0	100
37	Amyl alcohol	−50	25
30	Aniline	0	130
23	Benzene	10	80
27	Benzyl alcohol	−20	30
10	Benzyl chloride	−30	30
49	Brine, (25% CaCl$_2$)	−40	20
51	Brine, (25% NaCl)	−40	20
44	Butyl alcohol	0	100
2	Carbon disulfide	−100	25
3	Carbon tetrachloride	10	60
8	Chlorobenzene	0	100
4	Chloroform	0	50
21	Decane	−80	25
6A	Dichloroethane	−30	60
5	Dichloromethane	40	50
15	Diphenyl	80	120
16	Diphenyl oxide	0	200
22	Diphenylmethane	30	100
16	Dowtherm A	0	200
24	Ethyl acetate	−50	25
42	Ethyl alcohol, 100%	30	80
46	Ethyl alcohol, 95%	20	80
50	Ethyl alcohol, 50%	20	80
1	Ethyl bromide	5	25
13	Ethyl chloride	−30	40
36	Ethyl ether	−100	25
7	Ethyl iodide	0	100
25	Ethylbenzene	0	100
39	Ethylene glycol	−40	200
2A	Freon 11 (CCl$_3$F)	−20	70
6	Freon 12 (CCl$_2$F$_2$)	−40	15
4A	Freon 21 (CHCl$_2$F)	−20	70
7A	Freon 22 (CHClF$_2$)	−20	60
3A	Freon 113 (CCl$_2$F–CClF$_2$)	−20	70
38	Glycerol	−40	20
28	Heptane	0	60
35	Hexane	−80	20

TABLE B.1.   Key for Figure B.1—*Continued*

No.	Liquid	Temperature Range (°C)	
		From	To
48	Hydrochloric acid 30%	20	100
41	Isoamyl alcohol	10	100
43	Isobutyl alcohol	0	100
47	Isopropyl alcohol	−20	50
31	Isopropyl ether	−80	20
40	Methyl alcohol	−40	20
13A	Methyl chloride	−80	20
14	Naphthalene	90	200
12	Nitrobenzene	0	100
34	Nonane	−50	25
33	Octane	−50	25
3	Perchlorethylene	−30	140
45	Propyl alcohol	−20	100
20	Pyridine	−50	25
11	Sulfur dioxide	−20	100
9	Sulfuric acid 98%	10	45
23	Toluene	0	60
53	Water	10	200
18	*meta*-Xylene	0	100
19	*ortho*-Xylene	0	100
17	*para*-Xylene	0	100

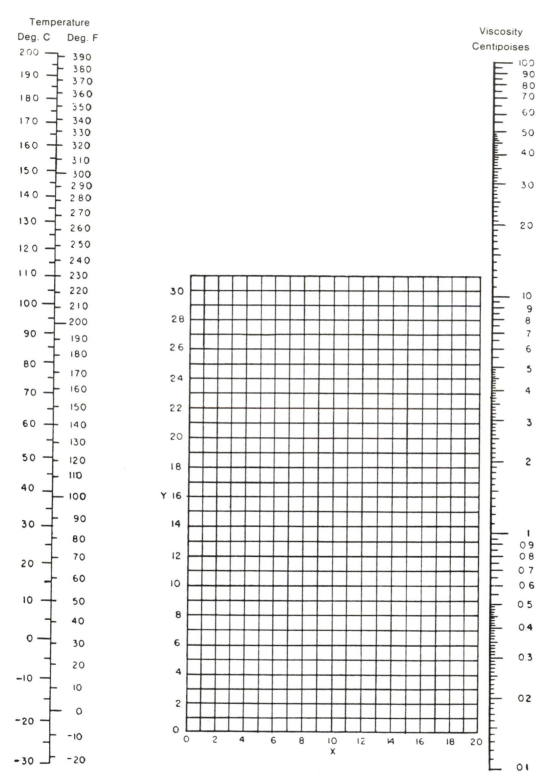

Temperature
Deg. C   Deg. F

Viscosity
Centipoises

X

Y

FIGURE B.2. Liquid viscosities. (Reproduced with permission from *Heat Transmission;* McAdams, W. H., Ed.; McGraw-Hill: New York, 1954. Copyright 1954 McGraw-Hill.)

TABLE B.2. X–Y Data for Use in Nomograph

Liquid	X	Y	Liquid	X	Y
Acetaldehyde	15.2	4.8	Diethyl ketone	13.5	9.2
Acetic acid, 100%	12.1	14.2	Diethyl oxalate	11.0	16.4
Acetic acid, 70%	9.5	17.0	Diethylene glycol	5.0	24.7
Acetic anhydride	12.7	12.8	Diphenyl	12.0	18.3
Acetone, 100%	14.5	7.2	Dipropyl ether	13.2	8.6
Acetone, 35%	7.9	15.0	Dipropyl oxalate	10.3	17.7
Acetonitrile	14.4	7.4	Ethyl acetate	13.7	9.1
Acrylic acid	12.3	13.9	Ethyl acrylate	12.7	10.4
Allyl alcohol	10.2	14.3	Ethyl alcohol, 100%	10.5	13.8
Allyl bromide	14.4	9.6	Ethyl alcohol, 95%	9.8	14.3
Allyl iodide	14.0	11.7	Ethyl alcohol, 40%	6.5	16.6
Ammonia, 100%	12.6	2.0	Ethyl bromide	14.5	8.1
Ammonia, 26%	10.1	13.9	Ethyl chloride	14.8	6.0
Amyl acetate	11.8	12.5	Ethyl ether	14.5	5.3
Amyl alcohol	7.5	18.4	Ethyl formate	14.2	8.4
Aniline	8.1	18.7	Ethyl iodide	14.7	10.3
Anisole	12.3	13.5	Ethyl propionate	13.2	9.9
Arsenic trichloride	13.9	14.5	Ethyl propyl ether	14.0	7.0
Benzene	12.5	10.9	Ethyl sulfide	13.8	8.9
Brine, 25% CaCL$_2$	6.6	15.9	Ethylbenzene	13.2	11.5
Brine, 25% NaCl	10.2	16.6	2-Ethylbutyl		
Bromine	14.2	13.2	acrylate	11.2	14.0
Bromotoluene	20.0	15.9	Ethylene bromide	11.9	15.7
Butyl acetate	12.3	11.0	Ethylene chloride	12.7	12.2
Butyl acrylate	11.5	12.6	Ethylene glycol	6.0	23.6
Butyl alcohol	8.6	17.2	2-Ethylhexyl		
Butyric acid	12.1	15.3	acrylate	9.0	15.0
Carbon dioxide	11.6	0.3	Ethylidene chloride	14.1	8.7
Carbon disulfide	16.1	7.5		13.7	10.4
Carbon tetrachloride	12.7	13.1	Formic acid	10.7	15.8
Chlorobenzene	12.3	12.4	Freon 11	14.4	9.0
Chloroform	14.4	10.2	Freon 12	16.8	15.6
Chlorosulfonic acid	11.2	18.1	Freon 21	15.7	7.5
meta-Chlorotoluene	13.3	12.5	Freon 22	17.2	4.7
ortho-Chlorotoluene	13.0	13.3	Freon 113	12.5	11.4
para-Chlorotoluene	13.3	12.5	Glycerol, 100%	2.0	30.0
meta-Cresol	2.5	20.8	Glycerol, 50%	6.9	19.6
Cyclohexane	9.8	12.9	Heptane	14.1	8.4
Cyclohexanol	2.9	24.3	Hexane	14.7	7.0
Dibromomethane	12.7	15.8	Hydrochloric acid,		
Dichloroethane	13.2	12.2	31.5%	13.0	16.6
Dichloromethane	14.6	8.9	Iodobenzene	12.8	15.9

TABLE B.2. *X–Y* Data for Use in Nomograph—*Continued*

Liquid	X	Y	Liquid	X	Y
Isobutyl alcohol	7.1	18.0	Phenol	6.9	20.8
Isobutyric acid	12.2	14.4	Phosphorus tribromide	13.8	16.7
Isopropyl alcohol	8.2	16.0	Phosphorus trichloride	16.2	10.9
Isopropyl bromide	14.1	9.2	Propionic acid	12.8	13.8
Isopropyl chloride	13.9	7.1	Propyl acetate	13.1	10.3
Isopropyl iodide	13.7	11.2	Propyl alcohol	9.1	16.5
Kerosene	10.2	16.9	Propyl bromide	14.5	9.6
Linseed oil, raw	7.5	27.2	Propyl chloride	14.4	7.5
Mercury	18.4	16.4	Propyl formate	13.1	9.7
Methanol, 100%	12.4	10.5	Propyl iodide	14.1	11.6
Methanol, 90%	12.3	11.8	Sodium	16.4	13.9
Methanol, 40%	7.8	15.5	Sodium hydroxide, 50%	3.2	25.8
Methyl acetate	14.2	8.2	Stannic chloride	13.5	12.8
Methyl acrylate	13.0	9.5	Succinonitrile	10.1	20.8
Methyl *n*-butyrate	13.2	10.3	Sulfur dioxide	15.2	7.1
Methyl chloride	15.0	3.8	Sulfuric acid, 110%	7.2	27.4
Methyl ethyl ketone	13.9	8.6	Sulfuric acid, 100%	8.0	25.1
Methyl formate	14.2	7.5	Sulfuric acid, 98%	7.0	24.8
Methyl iodide	14.3	9.3	Sulfuric acid, 60%	10.2	21.3
Methyl isobutyrate	12.3	9.7	Sulfuryl chloride	15.2	12.4
Methyl propionate	13.5	9.0	Tetrachloroethane	11.9	15.7
Methyl propyl ketone	14.3	9.5	Thiophene	13.2	11.0
Methyl sulfide	15.3	6.4	Titanium tetrachloride	14.4	12.3
Naphthalene	7.9	18.1	Toluene	13.7	10.4
Nitric acid, 95%	12.8	13.8	Trichloroethylene	14.8	10.5
Nitric acid, 60%	10.8	17.0	Triethylene glycol	4.7	24.8
Nitrobenzene	10.6	16.2	Turpentine	11.5	14.9
Nitrogen dioxide	12.9	8.6	Vinyl acetate	14.0	8.8
Nitrotoluene	11.0	17.0	Vinyl toluene	13.4	12.0
Octane	13.7	10.0	Water	10.2	13.0
Octyl alcohol	6.6	21.1	*meta*-Xylene	13.9	10.6
Pentachloroethane	10.9	17.3	*ortho*-Xylene	13.5	12.1
Pentane	14.9	5.2	*para*-Xylene	13.9	10.9

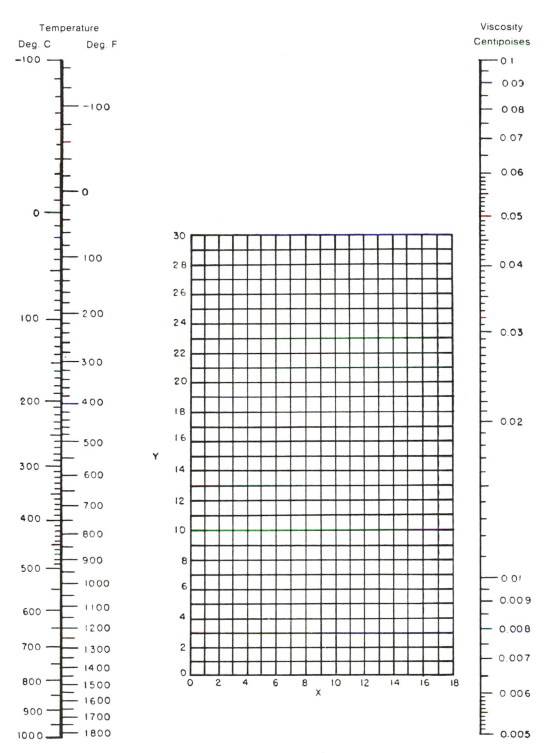

FIGURE B.3. Gas viscosities. (Reproduced with permission from *Heat Transfer;* McAdams, W. H., Ed.; McGraw-Hill: New York, 1954. Copyright 1954 McGraw-Hill.)

TABLE B.3. *X–Y* Data for Use in Nomograph

No.	Gas	X	Y	No.	Gas	X	Y
1	Acetic acid	7.7	14.3	29	Freon 113	11.3	14.0
2	Acetone	8.9	13.0	30	$3H_2(+)4N_2$	11.2	17.2
3	Acetylene	9.8	14.9	31	Helium	10.9	20.5
4	Air	11.0	20.0	32	Hexane	8.6	11.8
5	Ammonia	8.4	16.0	33	Hydrogen	11.2	12.4
6	Argon	10.5	22.4	34	Hydrogen bromide	8.8	20.9
7	Benzene	8.5	13.2	35	Hydrogen chloride	8.8	18.7
8	Bromine	8.9	19.2	36	Hydrogen cyanide	9.8	11.9
9	Butene	9.2	13.7	37	Hydrogen iodide	9.0	21.3
10	Butylene	8.9	13.0	38	Hydrogen sulfide	8.6	18.0
11	Carbon dioxide	9.5	18.7	39	Iodine	9.0	18.1
12	Carbon disulfide	8.0	16.0	40	Mercury	5.3	22.9
13	Carbon monoxide	11.0	20.0	41	Methane	9.9	15.5
14	Chlorine	9.0	18.4	42	Methyl alcohol	8.5	15.6
15	Chloroform	8.9	15.7	43	Nitric oxide	10.9	20.5
16	Cyanogen	9.2	15.2	44	Nitrogen	10.6	20.0
17	Cyclohexane	9.2	12.0	45	Nitronyl chloride	8.0	17.6
18	Ethane	9.1	14.5	46	Nitrous oxide	8.8	19.0
19	Ethyl acetate	8.5	13.2	47	Oxygen	11.0	21.3
20	Ethyl alcohol	9.2	14.2	48	Pentane	7.0	12.8
21	Ethyl chloride	8.5	15.6	49	Propane	9.7	12.9
22	Ethyl ether	8.9	13.0	50	Propyl alcohol	8.4	13.4
23	Ethylene	9.5	15.1	51	Propylene	9.0	13.8
24	Fluorine	7.3	23.8	52	Sulfur dioxide	9.6	17.0
25	Freon 11	10.6	15.1	53	Toluene	8.6	12.4
26	Freon 12	11.1	16.0	54	2,3,3-Trimethyl-		
27	Freon 21	10.8	15.3		butane	9.5	10.5
28	Freon 22	10.1	17.0	55	Water	8.0	16.0
				56	Xenon	9.3	23.0

TABLE B.4. Prandtl Numbers for Gases at 1 atm and 212 °F

Gas	$C_P\mu/k$
Air, hydrogen	0.69
Ammonia	0.86
Argon	0.66
Carbon dioxide, methane	0.75
Carbon monoxide	0.72
Helium	0.71
Nitric oxide, nitrous oxide	0.72
Nitrogen, oxygen	0.70
Steam (low pressure)	1.06

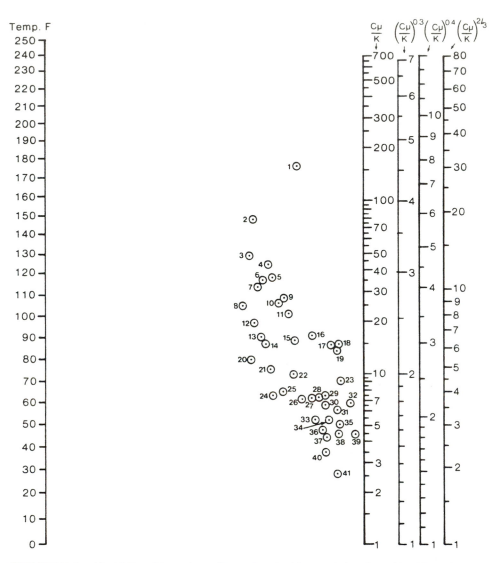

FIGURE B.4. Liquid Prandtl numbers. (Reproduced with permission from *Heat Transmission;* McAdams, W. H., Ed.; McGraw-Hill: New York, 1954. Copyright 1954 McGraw-Hill.)

TABLE B.5. Key for Figure B.4

No.	Liquid	Range (°F)
19	Acetic acid, 100%	50–140
14	Acetic acid, 50%	50–194
36	Acetone	14–176
21	Ammonia, 26%	14–230
18	Amyl acetate	32–104
6	Amyl alcohol	86–212
7	Aniline	14–248
28	Benzene	32–194
25	Brine, 25% $CaCl_2$	−4–176
20	Brine, 25% NaCl	−4–122
9	Butyl alcohol	14–230
41	Carbon disulfide	14–212
29	Carbon tetrachloride	32–176
23	Chlorobenzene	32–194
33	Chloroform	32–176
32	Ethyl acetate	32–140
17	Ethyl alcohol, 100%	14–212
16	Ethyl alcohol, 95%	50–158
12	Ethyl alcohol, 50%	50–176
40	Ethyl bromide	14–104
37	Ethyl ether	14–158
39	Ethyl iodide	14–176
1	Ethylene glycol	32–122
6	Glycerol, 50%	32–158
31	Heptane	14–140
34	Hexane	68–140
22	Hydrochloric acid, 30%	50–176
5	Isoamyl alcohol	50–230
10	Isopropyl alcohol	32–212
27	Methyl alcohol, 100%	14–176
13	Methyl alcohol, 40%	14–176
15	Nitrobenzene	68–212
35	Octane	14–122
36	Pentane	14–122
11	Propyl alcohol	86–176
2	Sulfuric acid, 111%	63–176
3	Sulfuric acid, 98%	50–194
4	Sulfuric acid, 60%	50–212
30	Toluene	14–230
24	Water	50–212
26	Xylene	14–122

Appendix C

Symbols

A	Arrhenius frequency factor
A	Component, as in binary mixtures
A	Cross-sectional area (ft^2, m^2)
A	Value used to determine constants required for C_d
A_e	Acceleration of a particle (ft/s^2, m/s^2)
A_o	Constant for Benedict–Webb–Rubin equation
A_{12}, A_{21}	Constants for vapor–liquid equations
a	Acceleration (ft/s^2, m/s^2)
a	Activity
a	Constant required for power number equation
a	Van der Waals constant
B	Bottoms product in mass-transfer operations (mol/h)
B	Component, as in binary mixtures
B	Redlich–Kister equation constant
B_o	Benedict–Webb–Rubin equation constant
b	Benedict–Webb–Rubin equation constant
b	Constant required for power number equation
b_1	Constant used for determining C_d
C	Constant for equation for heat transfer with flow over cylinders
C	Controlled variable
C	Redlich–Kister equation constant
C'	Empirical constant for nucleate boiling equation
C_A	Concentration of component A (lb-mol/ft^3, g-mol/m^3)
C_d	Drag coefficient
C_v	Coefficient of discharge
C_1	Constant
C_o	Benedict–Webb–Rubin equation constant
C_r	Replacement cost
C_{liquid}	Specific heat of liquid (Btu/lb-°F, J/kg-°C)

C_P	Specific heat at constant pressure (Btu/lb-°F, J/kg-°C)
C_V	Specific heat at constant volume (But/lb-°F, J/kg-°C)
C_{vapor}	Specific heat of vapor (Btu/lb-°F, J/kg-°C)
C_e	Specific heat at constant electromotive force
C_q	Specific heat at constant charge
D	Component in chemical reaction
D	Derivative
D	Differential
D	Diameter of a tube (ft, m)
D	Overhead product in mass-transfer operations (mol/h)
D'	Agitator diameter (ft, m)
D_{AB}, D_{BA}	Diffusion coefficient of component A through B (and the opposite) (ft²/h, m²/s)
D_L	Longitudinal diffusivity (ft²/h, m²/s)
D_p	Particle diameter (ft, m)
D_t	Tank diameter (ft, m)
E	Arrhenius activation energy (kcal/mol)
E	Clearance between impeller and tank (ft)
E	Component in chemical reaction
E	Energy (J, Btu)
E	Energy of activation (kcal/mol)
$\underline{E}$	Molar mass velocity
$\overline{E}$	Electric field
E_b	Energy radiated from a black body (Btu/h, J/s)
e	Electromotive force, electromagnetic force
e	Error
F	Component in chemical reaction
F	Degrees of freedom in Gibbs phase rule
F	Feed to a mass-transfer operation (kg-mol/h, lb-mol/h)
F	Force (lbf, N)
$\underline{F}$	Gibbs free energy (J, Btu)
$\overline{F}_{excess}$	Excess Gibbs free energy (J, Btu)
F	Shape or view factor in radiation
F_B	Buoyant force (lbf, N)
F_h	Friction heating term (Btu/lb-mass, J/kg)
F_{mn}	Fraction of energy transferred from surface m to surface n in radiative heat transfer
F_{12}, F_{21}	Fraction of energy transferred from surface 1 to surface 2 (and vice versa)
Fr	Froude number, dimensionless comparison between inertial and gravitational forces
$F°$	Standard Gibbs free energy (J, Btu)
f	Empirical friction factor
f	Fanning friction factor
f	Feed condition
f	"Function of"
f	Fugacity (N/m², lbf/in.²)
f_i	Fugacity of pure component i
f	Subscript indicating relation to a film
f^*	Modified friction factor

f'	Empirical factor for pressure drop over tube banks
f°	Standard-state fugacity (N/m^2, lbf/in.^{2})
$\bar{f}_i, \bar{f}_1, \bar{f}_2$	Fugacity of component i, or of a component in phase 1 or 2 in a solution
$\bar{f}_i^L, \bar{f}_i^V$	Fugacity in the liquid or vapor phase of a solution of component i
$f_{i,P*}^L, f_{i,P*}^V$	Fugacity in the liquid or the vapor of component i at partial pressure P*
f_i^0, f_1^0, f_2^0	Fugacity of component i, or of a component in phase 1 or 2 of a solution at standard state
G	Mass velocity (lbm/ft^2-h, kg/m^2-s)
$G_m, G_{m'}$	Molar mass velocity
Gr	Grashof number
$G(p)$	Transfer function
g	Local acceleration due to gravity (ft/s^2, m/s^2)
g_c	Universal constant: conversion factor (lbm-ft/s^2-lbf, kg-m/s^2-N)
H	Depth of liquid in pressure and flow calculations
H	Enthalpy (J, Btu)
Ha	Hatta number, dimensionless comparison of mass transfer with and without chemical reaction (Chapter 6)
$\Delta H_{reaction}$	Heat of reaction (Btu/lb-mol, J/kg-mol)
$\bar{H}$	Magnetic intensity
H_{OG}, H_{OL}, H_L, H_G	Packing height (m, ft)
h	Distance between plates (in., cm)
h	Heat-transfer coefficient (Btu/h-ft^2-°F, W/m^2-°C)
h	Height (m, ft)
$\bar{h}$	Molar enthalpy (J/mol, Btu/lb-mol)
$\bar{h}^0$	Molal enthalpy at standard state
h_m	Mean heat transfer coefficient (Btu/h-ft^2-°F, W/m^2-°C)
I	Integral
i	Interest rate
i	Unit vector in x direction
J	Molar flux vector (lb-mol/s-ft^2)
J_A	Mass-transfer flux with no volume transport across a plane (lb-mol/s-ft^2)
j	Unit vector in y direction
j_D	Mass-transfer flux
j_H	Heat-transfer flux
K	Consistency index (lbf-s^2/ft^2)
K	Equilibrium constant
K	Power law constant
K	Proportionality constant for expansion and contraction
K	Resistance coefficient for fluid flow calculations
K_s	Sensitivity or gain for process controller
K^*, K^{**}	Equilibrium constant for adsorption–reaction–desorption system
k	Specific reaction rate
k	Thermal conductivity (Btu/h-ft^2-°F, W/m-°C)

k	Unit vector in the z direction
$k*$	Mass-transfer coefficient (lb-mol/ft^2-h, g-mol/m^2-s)
k_c	Mass-transfer coefficient (ft/h, m/s)
k_g	Mass-transfer coefficient (lb-mol/h-ft^2-atm, g-mol/s-N/m^2)
k_p	Specific reaction rate constant (m^2/(N $\cdot$ s))
k_p	Mass-transfer coefficient (lb-mol/h-ft^2, g-mol/s-m^2)
k_y	Mass-transfer coefficient (lb-mol/h-ft^2, g-mol/s-m^2)
L	Length (m, ft)
L	Thickness of filter (ft, m)
L_a	Liquid rate in distillation column enriching section (lb-mol/h, g-mol/s)
L_m, L'_m	Liquid rate in distillation column stripping section (lb-mol/h, g-mol/s)
L_{m-1}	Liquid rate in distillation column stripping section (lb-mol/h, g-mol/s)
L_n	Liquid rate in distillation column rectifying section (lb-mol/h, g-mol/s)
L_{n-1}	Liquid rate in distillation column rectifying section (lb-mol/h, g-mol/s)
$\underline{l}$	Length in a particular direction (ft, m)
M	Control output
m	Constant used for determining C_d
m	Mass
m	Ratio of k/hx_1
$\dot{m}$	Mass flow rate (lb/h, m/s)
N	Agitator speed (rpm)
N	Number of transfer units
N	Number of transverse rows of tubes in a heat exchanger
N_A	Molar mass flux (lb-mol/ft^2-h, g-mol/m^2-s)
N_i	Mass flux of the ith component (lb-mol/ft^2-h, g-mol/m^2-s)
Nu	Nusselt number, dimensionless comparison of inertial forces and heat transfer by convection, and viscous forces and heat transfer by conduction
n	Dimensionless factor, group, or position
n	Exponent for equation for heat transfer with flow over cylinders
n	Exponent for power law
n	Flow behavior index
n	Number of moles of substance
n	Years, used in engineering economic calculations
P	Phases
P_o	Power number, dimensionless
P	Pressure (lbf-in.^{2}, N/m^2)
P	Present worth, used in engineering economic calculations
P_{BM}	Bulk mean pressure (lbf/ft^2)
P_c	Critical pressure of a substance
P_c'	Pseudocritical pressure of a mixture of gases; equal to the sum of each component's mole fraction in the gas multiplied times its critical pressure

P_E	Molar mass velocity of product (lb-mol/ft^2-h, kg-mol/m^2-s)
P_R	Reduced pressure
P'_R	P/P_c'
P^*	Base pressure, generally at 0 °C and atmospheric pressure
Pr	Prandtl number, dimensionless comparison of momentum and thermal diffusivities
$\overline{P}$	Total polarization
P^0	Pressure at standard state
P'_c	Pressure at critical state
Q	Heat energy (Btu/lb-mass, J/kg)
Q	Net molar flow rate (lb-mol/h, kg-mol/s)
Q	Volumetric flow rate (ft^3/s, m^3/s)
$\dot{Q}$	Heat-transfer rate, dQ/dt
q	Charge
q	Heat flux (Btu/h-ft^2, W/m^2)
q	Heat from an internal source
q_x, q_y, q_z	Heat flux in x, y, and z direction respectively (Btu/h-ft)
R	Annual payment, used in engineering economic calculations
R	Electrical resistance (Ω)
R	Gas law constant (Btu/lb-mol-°R, J/g-mol-K)
R	Reflux ratio for mass-transfer equipment calculations
R_A	Rate of chemical reaction
R_n	Hydraulic radius (ft, m)
R_m	Fluid column length in a manometer (ft, m)
R_o	Molar mass velocity of recycle stream (lb-mol/h, kg-mol/s)
Re	Reynolds number, dimensionless comparison of inertial and viscous forces in fluid flow
Re$_p$	Reynolds number for a particle
r	Radius (ft, m)
r	Reaction rate (lb-mol/h, g-mol/s)
S	Entropy (J/K)
$\overline{S}$	Average entropy of a flowing fluid (J/K)
S	Future sum of money, used in engineering economic calculations
S	Projected area perpendicular to flow (ft^2, m^2)
S_E	Molar mass velocity of recovered solvent
S_n	Distance separating the center lines of two adjacent vertical pipes (ft, m)
S_n	Normal spacing of tube banks (ft, m)
S_o	Specific particle area (ft^2, m^2)
S_p	Distance separating the center lines of two adjacent horizontal pipes (ft, m)
S_p	Parallel spacing of tube bank (ft, m)
Sc	Schmidt number, dimensionless comparison of mass transfer by convection and diffusion
s	Entropy per mass (J/kg-K, Btu/lb-mass-°R)
$\bar{s}$	Molar entropy (J/kg-mol-K, Btu/lb-mol-°R)
T	Temperature (°C, °F, K, °R)
T_b	Temperature at the bubble point of a liquid (°C, °F, K, °R)

T_c	Critical temperature (K, °R)
T_c'	Pseudocritical temperature of a mixture of gases equal to the sum of the mole fraction of each component i times its critical temperature, $T_c' = \sum y_i (T_c)i$
T_d	Temperature at the dew point (°C, °F, K, °R)
T_F	Temperature at the feed point in a mass-transfer unit such as a distillation column (°C, °F, K, °R)
T_R	Reduced temperature
T_R'	T/T_c'
T^*	Base temperature (usually in °C)
t	Time (h, s)
t_T	Blending time (h, s)
t_w	Temperature at wall, used for heat-transfer calculations (°C, °F, K, °R)
U	Internal energy (Btu/lb-mol)
U	Overall heat-transfer coefficient (Btu/ft²-h-°F, W/m²-K)
U_y	Velocity of mass-transfer boundary (ft/s, m/s)
U_y	Velocity of a moving plane with mass transfer (ft/s, m/s)
u	Internal energy (J, Btu)
$\bar{u}$	Molar internal energy (J-kg-mol, Btu-lb-mol)
V	Velocity of fluid flow (ft/s, m/s)
$\bar{V}$	Average velocity (ft/s, m/s)
V_s	Superficial velocity (ft/s, m/s)
v	Volume (ft³, m³)
v_i	Partial molal volume of ith component (ft³/lb-mol, m³/kg-mol)
$\bar{v}$	Molar volume (ft³/lb-mol, m³/kg-mol)
$V_a, V_b, V_m,$ $V_{m+1}, V_n,$ V_{n+1}, V_{n-1}	Vapor rate leaving stage a, b, m, $m + 1$, n, $n + 1$, $n - 1$, respectively (kg-mol/h)
v_s	Salvage value, used in economic calculations
vol_m	Volume of solid–liquid suspension (ft³, m³)
W	Baffle width (ft, m)
W	Net molar flow rate (kg-mol/h)
W	Work (ft-lbf, J)
W_A	Molar rate of exchange of A (mol/h)
W_s	Shaft work (Btu/lb-mass, J/kg)
$\dot{W}_s$	Shaft work per time (Btu/lb-mass-s, J/kg-s)
X	Coordinate value for unsteady-state heat conduction graph
X	Distance in flow direction (ft, m)
X	Liquid mole fraction
X'	Ratio of mole fraction in liquid to mole fraction in vapor
X_n	Mole fraction in liquid leaving stage n
X_T	Plate length (ft, m)
x	Coordinate distance (ft, m)
x	Exponent for agitator power equation
x	Thickness (ft, m)
x_A	Mole fraction of component A
x_B, x_F, x_D	Mole fraction of the tie component in bottoms, feed, and distillate streams, respectively
x_1	Half thickness of slab (ft, m)

Y	Reduced temperature
Y	Dimension ratio
Y_A, Y_B	Mole fraction in vapor of component A and B, respectively
y	Coordinate distance (ft, m)
y	Mole fraction of vapor
y_A	Mole fraction of component A
y_{BM}	Logarithmic mole fraction in vapor to mole fraction in liquid
y_n	Mole fraction in vapor leaving state n
y_0	Film thickness (ft, m)
y_0	Slab thickness (ft, m)
y'	Ratio of mole fraction in vapor to mole fraction in liquid
Z	Dimension ratio, dimensionless
Z	Height above a datum plane for fluid flow calculations (ft, m)
Z_i	Agitator elevation (ft, m)
Z_1	Height of liquid in agitated tank (ft, m)
Z_m	Length of fluid column in manometer (ft, m)
Z_s	Height of suspension in mixing system (ft, m)
z	Height (ft, m)

GREEK

α	Absorptivity, dimensionless
α	Benedict–Webb–Rubin equation constant
β	Calculated exponent for agitator power equation
β	Coefficient of thermal expansion, $1/°R$, $1/K$
γ	Surface tension
γ_i	Activity coefficient of pure component i
γ_1, γ_2	Activity coefficient in phase 1 or 2
$\gamma_1^\infty, \gamma_2^\infty, \gamma_i^\infty$	Activity coefficient of pure component i or i in phase 1 or 2 of a solution as the mole fraction of i as the liquid approaches zero
γ_i^L, γ_i^v	Activity coefficient of component i in the liquid or vapor phase of a solution
Γ	Liquid feed per unit width (lbm/ft-h, kg/m-s)
Γ	Mass flow rate per unit perimeter (lbm/ft-h, kg/m-s)
ϵ	Emissivity, dimensionless
ϵ	Porosity or fraction of total volume that is void
ϵ	Size of roughness
ϵ	Strain
ϵ_m	Volume fraction of liquid in a suspension
θ	Angle
θ	Fraction of surface covered
θ	Time (h, s)
λ	Heat of vaporization (Btu/lbm, J/kg)
μ	Chemical potential
μ	Joule–Thomson coefficient
μ	Viscosity (lbm/ft-h, kg/m-s)
μ_{app}	Apparent viscosity (lbm/ft-h, kg/m-s)
μ_b	Bulk viscosity (lbm/ft-h, kg/m-s)
μ_i	Chemical potential of component i
μ_i^0	Chemical potential of component i at standard state

μ_i^L, μ_i^v	Chemical potential of component i in the liquid or vapor phase of a multicomponent mix
μ_w	Viscosity of a fluid at a wall (lbm/ft-h, kg/m-s)
ρ	Density (lbm/ft^3, kg/m^3)
ρ_b	Bulk density (lbm/ft^3/ kg/m^3)
ρ_{fluid}	Fluid density (lbm/ft^3, kg/m^3)
ρ_g	Gas density (lbm/ft^3, kg/m^3)
ρ_l	Liquid density (lbm/ft^3, kg/m^3)
ρ_m	Density of a solid–liquid suspension (lbm/ft^3, kg/m^3)
$\rho_{particle}$	Particle density (lbm/ft^3, kg/m^3)
Σ	Cross-sectional area of a sedimentation unit matching a given centrifuge (ft^2, m^2)
σ	Stefan–Boltzmann constant (Btu/h-ft^2-°R^4, W/m^2-K)
σ	Stress (lbf/ft^2, N/m^2)
τ	Shear stress (lbf/ft^2, N/m^2)
τ	Time constant
ϕ	Function
ϕ_i	Fugacity coefficient for pure component i; $\phi_i = f_i/P$
ϕ_i	Fugacity coefficient for component i in solution $\phi_i = \dfrac{f_i}{x_i P_i}$
ψ	Function
ζ	Damping coefficient

Index

A

Absorption
    factor, 184–188
    packed column, 181–189
    plate column, 179–181
Acid–base catalysis, 221
Activation energy, 7
Activity
    defined, 53
    fugacity relation, 207–208
Activity coefficient
    defined, 55
    equations with two constants, 64–66
    for vapor general form, 62
Adaptive control, 279
Adiabatic condition
    defined, 8
    operation, 8–9
Adiabatic reactor
    defined, 223–224
    isothermal, 227–229
Apparent viscosity, 98
Arrhenius equation, 7, 209, 225

B

Barometric equation, 69
Base pressure, 34
Base temperature, 34, 223
Batch distillation, 177–178
Batch reactor, 8, 9–10, 12–13, 234
Benedict–Webb–Rubin equation, 56
Bernoulli's equation, 74–80, 89
Bode plot, frequency response, 275–276

Boiling
    nucleate, 130–131
    points, 164–165, 170–180
    stages, 130
Boundaries
    defined, 2
    multiple, 4
    phase, 154–155
Buoyancy, 72

C

Capital investment cost, 296–299
Capital recovery factor, 284
Carnot cycle, 44–46
Catalytic cracker, 11, 18
Catalyzed reaction, 16–18, 231–234
Centrifugal pump, 91
Centrifugation, 198–201
Centrifuges, 196–197, 199–201
Chemical Engineering plant index, 292–293
Chemical equilibrium, 205–208
Chemical potential
    defined, 52
    fugacity relation, 207
Coefficient
    activity, 55, 62–66
    discharge, 78
    distribution, 54, 56, 60–61
    drag, 96–97, 199
    fugacity, 57
    heat transfer, 118–124
    Joule–Thomson, 34
    mass transfer, 154–155, 161, 231
    performance, 50